LITES - Legal Issues in Transdisciplinary Environmental Studies

Volume 5

The 'Legal Issues in Transdisciplinary Environmental Studies' (LITES) Book Series is based on the assumption that the process of dialogue and cultural integration between law, life and earth sciences, and social and human sciences should be strengthened and updated, by relying on transdisciplinary research platforms such as agroecology, environmental studies, environmental science, and sustainability science. According to the new paradigm of social-ecological systems (SES), the concept of the environment is conceived as a complex system of relationships between ecological and social factors, including the cultural and economic ones. The primary purpose of law, in this conceptual framework, is to preside over the durability of the essential conditions for the survival of the social-ecological systems and the protection of life at all scales (of individuals, societies, ecosystems).

LITES Series aims to explore the relationships between legal and environmental sciences according to a transdisciplinary perspective. On the one hand, natural and social environmental sciences need to integrate the point of view of law: this entails to study the complexities of SES in the light of normative and institutional variables, with the lens of categories such as rights, duties, powers, responsibilities, and procedural safeguards. On the other hand, law is called upon to review its own internal geometries, confronting them with the holistic approach toward sustainability in the scientific debate. Accordingly, law should address the need of changing the approach that so far has led to both hypertrophy and disarticulation when regulating closely linked matters such as the environment, agriculture, forestry, landscape and cultural heritage, energy, and food.

LITES Series is addressed to a wide international and interdisciplinary readership, targeting academic researchers and scholars, experts and practitioner lawyers, public administrations, judges, and law-makers. Its volume editors and contributing authors have different backgrounds and come from all over the world in order to provide a forum for discussion and normative analysis about new legal frontiers of human-environment interactions across disciplinary barriers.

Sadiya S. Silvee

The Law of Food Security: The Legal Framework for a Sustainable and Equitable Global Food System

Springer

Sadiya S. Silvee
School of Law and School of History
Zhengzhou University
Zhengzhou, Henan, China

ISSN 2522-5049 ISSN 2522-5057 (electronic)
LITES - Legal Issues in Transdisciplinary Environmental Studies
ISBN 978-981-95-3635-1 ISBN 978-981-95-3636-8 (eBook)
https://doi.org/10.1007/978-981-95-3636-8

This Springer imprint is published by the registered company Springer Nature Singapore Pte Ltd.
The registered company address is: 152 Beach Road, #21-01/04 Gateway East, Singapore 189721, Singapore

This book is dedicated to my parents, Late. Bir. Khondaker Ataul Haque & Ms. Rashida Banu for their unwavering support, and constant encouragement

Preface

Food security, hunger, and malnutrition are long-term issues that have a global impact on food security, resulting in an increase in the number of hungry people from 811 million in 2018 to 820 million in 2020. In an age of unprecedented agricultural abundance, it is a profound and tragic paradox that global hunger is on the rise. While our farms produce more than enough to feed everyone on the planet, the number of people facing food insecurity continues to grow. The question of *why* our efforts fall short has been the most critical and difficult challenge of our time. This book confronts this paradox head-on, arguing that the problem is not a lack of food, but a fundamental failure of the system designed to govern it.

At the heart of this failure lies a flawed and inequitable international legal framework. Hence, it is necessary to analyze contemporary international food documents from both *legal and social perspectives*. From a *legal perspective,* it is important to consider the classification, sources, morality, necessity of laws, and the nature of the rule of law. From a *social perspective*, attention should be paid to the social attitudes, behaviors, meanings, and constraints of developed and developing countries. For example, regarding food waste, do countries share common values? Do countries worldwide have common ethical views?

This book reveals how the contemporary global food system, governed by a web of conflicting international laws, prioritizes the mechanics of trade and food safety over the fundamental human right to food. We demonstrate how these rules create a deep chasm between developed and developing nations, imposing stringent standards that act as trade barriers, promote waste, and ultimately deepen the very insecurity they should prevent. To uncover the roots of this systemic dysfunction, this study embarks on a multidisciplinary investigation. We journey through a century of international food law, conduct empirical research into the staggering scale of food waste, and analyze the complex interplay of legal, social, and economic forces that shape our global food supply.

This book culminates not in mere critique, but in a concrete blueprint for reform and puts forward a comprehensive solution: a new institutional framework for fair governance, a novel "Food Waste Severity Measuring Scale" to make a hidden problem visible, and the detailed outline of a new multilateral treaty—the "Law of Food Security." Together, these proposals offer a viable path toward transforming our broken system into one that is truly sustainable, equitable, and capable of feeding the world for generations to come.

This study adopts an empirical legal research method and utilizes a Mixed Methods Research (MMR) design, combining qualitative and quantitative data collection and analysis methods. It follows an exploratory sequential design, combining data collection and analysis in a series of stages. The research consists of two main phases: the first phase involves the use of systematic review and Critical Interpretive Synthesis (CIS) to analyze qualitative data, combining the Theorizing Regulatory Intermediaries (RIT) model and Institutional Analysis and Development (IAD) framework to analyze contemporary legal mechanisms, institutional frameworks, and food governance within the global food system. The second phase involves the use of meta-analysis and statistical analysis for quantitative data analysis to support the findings from the qualitative data analysis.

This study conducted a web-based survey at X University in China to assess people's food waste attitude at food service places such as, in restaurants. The survey questionnaire was divided into four parts: demographic information, dietary habits in restaurants, awareness of food waste as a global issue, and anti-food waste laws. The data were analyzed using the IBM Statistical Package for the Social Sciences (SPSS) software, employing "descriptive frequency analysis tests" and cumulative inference of results. The combination of qualitative and quantitative data collection and analysis in this study provides increased evidence support, making it an evidence-based study.

Chapter 1 of this book provides an overview of key themes and concepts in research on food security and food waste. It emphasizes the need to establish a sustainable and fair food system to address the challenges posed by climate change, unequal resource distribution, and global food governance. Building on this foundation, Chap. 2 focuses on the relationship between food insecurity, food security, and food safety, and analyzes the interconnections of these concepts in the global food system. By examining the three main concepts of "food insecurity," "food security," and "food safety," Chap. 2 provides a solid basis for understanding the role of food waste in the food system, which is also the focus of Chap. 3. Chapter 3 explores the relationship between food waste and food security, emphasizing the need to establish a sustainable and fair food system. Through meta-analysis and statistical analysis of previous research and web-based surveys, this chapter provides valuable insights into the impact of food waste on global food supply, the attitudes of different cultures and regions toward food waste, and the importance of addressing this issue in achieving food security. The quantitative data analysis in this chapter indicates that the average food waste in the global food supply chain is approximately 36% annually. Additionally, descriptive frequency analysis estimates that the rate of food waste in restaurants by individuals is $\geq$22.51.

Building on this, Chap. 4 examines the evolution of international food instruments from international alliances to the present day. By tracing the development of these instruments, this chapter reveals the historical background of food insecurity and the evolution of international response measures, from the post-World War II focus on food security to the current emphasis on food quality, standards, and waste. This historical perspective provides a basis for understanding the current international food instruments, the state of global food governance, and the challenges faced in the development of a sustainable and equitable food system discussed in Chap. 5. Based on the arguments and conclusions presented in this paper, Chap. 6 outlines the *Law of Food Security* with 14 provisions. This outline is developed by analyzing and considering the international and national food, agricultural, food and nutritional policies, and legislation of twenty (20) developing and developed countries, including China, Bangladesh, India, and so forth. Implementation at the national and international levels would require the support of the United Nations and its related agencies, as well as cooperation from countries through ratification or accession.

Most importantly, this book does not leave with a diagnosis of despair. Instead, it offers a tangible and hopeful blueprint for the future. The book presents concrete, actionable solutions that have the potential to reshape global policy. The proposal of a "Food Waste Severity Measuring Scale," the development of a new institutional model for the global food system, and the detailed outline of a multilateral "Law of Food Security" treaty are not just academic proposals; they are the essential building blocks for a more just and effective system.

This book is a call to action and an indispensable resource for policymakers, scholars, and advocates. It challenges us to rethink our assumptions and provides the tools to finally build a global food system that is fair, sustainable, and capable of fulfilling its most basic promise: to ensure no one goes hungry.

Zhengzhou, China Sadiya S. Silvee

Acknowledgment This study has been one of the most challenging experiences of my life. I am indebted to many people without whose constant material and emotional support and encouragement throughout the process, this study would have never been completed.

First, I would like to express my sincere gratitude to my supervisor Professor Dr. Ximei Wu, whose proper guidance, invaluable support, and encouragement helped me to complete this work which will indeed induce further research on this field. Secondly, my special thanks and appreciation go to Zeng Wen Ge, Dr. Chen Yongmei, Dr. Hu Xiao Hong, Dr. Jiang Shi Bo, Dr. Shi Mao Sheng, Dr. Lu Ming Yu, and Dr. Li Jian Xin, for providing me their insightful feedback. Their critical evaluation, attention to detail, and valuable suggestions have challenged me to think critically and to further develop my ideas.

Special thanks go to my mother, Ms. Rashida Banu for being there for me and always being supportive and encouraging throughout this process. Most of all greatest thanks goes to my Mother. She was constantly by my side and has shared the pain, stress, despair, and tears of this difficult journey. She provided the crucial emotional pillars of support for me to survive this process. She has been beside me in every page and every single line of this work. Besides, I would like to specially thank my husband, Mr. Madanoor Mohamed Wasaf for standing by my side and constantly boosting my energy for further work.

Finally, my deepest gratitude goes to my beloved father, Late. Bir. Khondaker Ataul Haque, whom I sadly lost. I find myself very unfortunate that he could not see the fruition of this work. He was always my inspiration and encouragement. This book is as much my parents as mine.

Competing Interests The author has no competing interests to declare that are relevant to the content of this manuscript.

Contents

Abbreviations

CAC	Codex Alimentarius Commission
CFS	Committee on World Food Security
CIS	Critical Interpretive Synthesis
CSA	Climate-Smart Agriculture
CSSD	Consultative Subcommittee on Surplus Disposal
FAO	Food Agricultural Organization
FIS	Food Insecurity
FLW	Food Loss and Waste
FS	Food Security
FSA	Food Security Agency
FSD	Food Security Department
GATT	General Agreement on Tariffs and Trade
GHG	Greenhouse Gas
HACCP	Hazard Analysis and Critical Control Points
IAD	Institutional Analysis and Development
ICC	International Criminal Court
ICCPR	International Covenant on Civil and Political Rights
ICESCR	International Covenant on Economic, Social and Cultural Rights
IFST	International Food Security Treaty
IHL	International Humanitarian Law
ILO	International Labour Organization
INFOSAN	International Food Safety Authorities Network
ITPGRFA	International Treaty on Plant Genetic Resources for Food and Agriculture
MDGs	Millennium Development Goals
MMR	Mixed Methods Research
OECD	Organization for Economic Cooperation and Development
RIT	Theorizing Regulatory Intermediaries
SDG	Sustainable Development Goals
SDT	Special and Differential Treatment

SOFI	State of Food Security and Nutrition in the World
SPS Agreement	Application of Sanitary and Phytosanitary Measures
TBT Agreement	Agreement on Technical Barriers to Trade
UDEHM	Universal Declaration on the Eradication of Hunger and Malnutrition
UDHR	Universal Declaration of Human Rights
UN	United Nation
UNDESA	United Nations Department of Economics and Social Affairs
UNDP	United Nations Development Programme
UNEP	United Nations Environment Program
UNFCCC	United Nations Framework Convention on Climate Change
UNICEF	United Nations Children's Fund
WEF	World Economic Forum
WFP	World Food Program
WHO	World Health Organization
WTO	World Trade Organization

Chapter 1
The Role of Law in Addressing Global Food Security Challenges

1.1 Introduction

Food security, a fundamental human right, is increasingly precarious in our complex world. It represents a multifaceted challenge demanding urgent attention and innovative solutions. While often perceived as a purely agricultural or economic issue, food security is intricately linked with legal frameworks, both at national and international levels. Food safety, hunger, malnutrition are lingering food insecurity problems impacting global food security since long, leading to more than 820 million hungry people in 2020 (FAO et al., 2021). The global community, both at international and national levels, plays a vital role in shaping the discourse and fortifying their relevance in food security policymaking. So, law plays a crucial role in shaping food systems, regulating access to resources, and mitigating the numerous factors threatening food security.

The concept of food security encompasses more than just the availability of food. It extends to ensuring access to sufficient, safe, and nutritious food for all individuals, regardless of their socio-economic status, geographic location, or other circumstances. This includes the physical and economic access to food, its proper utilization through adequate nutrition and sanitation, and the stability of food supplies over time. In a world grappling with climate change, population growth, and persistent inequalities, achieving and maintaining food security is a formidable task. Law, in its various forms, is an indispensable tool for addressing food security challenges. International human rights law recognizes the right to adequate food as a fundamental human right, placing obligations on states to respect, protect, and fulfill this right. International trade law, intellectual property law, and environmental law all have significant implications for food systems and access to resources. At the national level, land tenure laws, water rights laws, and agricultural policies play a crucial role in shaping food production and distribution.

However, the legal landscape surrounding food security is complex and often fragmented. Conflicting norms, overlapping jurisdictions, and gaps in legal frameworks

S. S. Silvee, *The Law of Food Security: The Legal Framework for a Sustainable and Equitable Global Food System*, LITES - Legal Issues in Transdisciplinary Environmental Studies 5, https://doi.org/10.1007/978-981-95-3636-8_1

can hinder effective action. Furthermore, the implementation and enforcement of existing laws remain a significant challenge in many parts of the world. Critical analysis of these legal frameworks is essential to identify weaknesses, promote coherence, and strengthen the legal foundations of food security.

This chapter delves into the intricate relationship between law and food security, exploring the various legal instruments and mechanisms that shape food systems and influence access to food. It examines the challenges and opportunities presented by existing legal frameworks, advocating for a more holistic and integrated approach to food security governance. By critically analyzing the role of law in addressing food security challenges, we can pave the way for a more just and sustainable food future for all.

This chapter is divided into eight (8) segments, including an introduction and conclusion. It begins by discussing the concepts of food security and food insecurity, then examines the global food system and its relationship with food security. The chapter further explores the drivers of food security and insecurity, as well as the issue of food waste, and the role of law in addressing food security. Finally, the chapter delves into the topic of global food governance.

1.2 Comprehending the Notion Food Security and Food Insecurity

Beginning with Reutlinger article titled as *Food security and poverty in LDCs* (Kang, 2015) where the author defined food security as "access by all people at all times to enough food for an active and healthy life." The author's this definition was influenced by the definition introduced in *1996 World Food Summit*[1]and *UN World Food Conference Report*, 1975,[2] and were reflected in the early literature on food security such as, Maxwell and Smith (1992), and Valdes (1981). However, in 1980s, Reutlinger and Knapp in the article *Food security in Food Deficit Countries* (Reutlinger & Knapp, 1980) defined food security, for the first time as, "a condition in which the probability of a country's citizens falling below a minimal level of food consumption is low." However, in both the literatures the level of food security in the least developing countries was explored and the focus of the study circulated around instability in food production, consumption, and prices. Furthermore, considering the nexus between food security and trade policies the author went on to explore the impact of trade policies and need to stabilize the trade policies to build a greater food security.

Holding the similar approach Ashley in his book *Food security in the Developing World* (Ashley, 2016) identifying the issue stated, food security is "too often

[1] Food security exists when all people, at all times, have physical and economic access to sufficient safe and nutritious food that meets their dietary needs and food preferences for an active and healthy life.

[2] Availability at all times of adequate world supplies of basic food-stuffs..., to sustain a steady expansion of food consumption [...] and to offset fluctuations in production and prices.

[…] equates to only *food availability*. Often missing or insufficient is imperative of seamless and holistic integration of food *availability* with economic, social and physical *access* of food and the *nutritional*, *utilization* and *food safety* components, as also the imperative that all these parameters of achieving food security need to be accounted for on a simultaneous, present continuous and stable basis. The author further defined food security as "the ability of a community, family or individual to be able to eat sufficiently, in terms of both quantity and quality, as prescribed by international standard […] of intake."

However, Baer-Nawrocka A. and Sadowski A gave a different approach in their work *Food security and food self-sufficiency around the world: A typology of countries* (Baer-Nawrocka & Sadowski, 2019) where they explored the current status of food security in different countries around the world, considering both aspects (physical and economic availability) combined together. The authors asserted that "food security depend on both natural and economic factors. […] international trade largely contributes to food security." Sumner defining food security in the article *Agricultural trade policy and food security* (Sumner, 2000) noted, "food security means more than a current absence of hunger or even the current possession of nutritional health". Food security concerns potential food intake into the future and is thus inherently dynamic". In addition, as implied by the word 'security' itself, any concept of food security must include the stochastic nature of food supply and demand. Besides the author has clearly demonstrated how trade impacts food security. For which he proposed five major channels: (1) income changes resulting from opening to trade, (2) impacts on food price volatility, (3) productivity gains from trade, and (4) changes in dietary diversity and quality.

On the other hand, Olper et al. in the working paper *Food Security, Health and Trade Liberalization* investigated the effect of trade liberalization on food security outcomes. He mainly focused on the macro determinants of food security and health outcomes, such as macro-economic shock, trade liberalization, and institutional and political reforms (Olper et al., 2014). Furthermore, Olper et al. in his later work *Trade liberalization and child mortality: A Synthetic Control Method* examining trade liberalization on children mortality, the authors also shed light on agricultural policy as he argued agricultural policy as one of the influential factors (Olper et al., 2018).

So, in the earlier literatures' food security is defined in terms of food availability and nutrition, reflection of which is visible in the present literature. Therefore, it can be stated that the notion food security is understood and defined as a situation where sufficient quantity of food is available and accessible by the people that can provide sufficient amount of essential nutrients. This means food security is the goal that we want to achieve because our problem is "food insecurity," which is acknowledged by authors of the above-mentioned literatures along with the other present literatures. But "food insecurity" is not well defined nor gained much attention, neither, the authors attempted to define "food security" as converse of "food insecurity." On the other hand, the definition often cited to define "food insecurity," states, "food insecurity as a situation where some people do not have access to sufficient quantities of safe and nutritious food and hence do not consume the food that they need to grow normally and conduct an active and healthy life."[3] The noticeable fact here is "food insecurity" and "food security" is defined as a situation. Although, "food security" cannot be measured, it is "food insecurity" which is measured, assessed, or analyzed to determine whether the situation can be addressed as "food security" or "food insecurity." However, the contemporary literatures, focused on "food security" than "food insecurity"; when both the notions are entangled with each other and on cannot be understood without the other. For instance, "good" cannot be understood without understanding "bad." So, the contemporary international food instruments need to give more emphasis on "food insecurity," which is the main problem, than "food security," which is the goal that needs to be achieved. And goal can be achieved only by emphasizing on the problem itself, rather than emphasizing on the goal.

1.3 Global Food System and Food Security

The global food system is a complex network of agricultural production, processing, distribution, and consumption. This system is heavily influenced by various legal frameworks, including international trade agreements, intellectual property rights, and environmental regulations. As the world has become more interconnected, the global food system has evolved significantly over time, playing a crucial role in shaping food security around the world. Historically, the transition from subsistence-based food production to globally interconnected food trade and distribution has transformed how people access and consume food (Clapp et al., 2022). This section discusses about the evolution of the global food system, globalization's impact on food security, and the role of trade policies, resource distribution, and market dynamics. Addressing these systemic issues is crucial for understanding the global food system.

1.3.1 The Evolution of Modern Human Societies and the Food Regime

The food system has undergone significant transformations as human societies have transitioned from subsistence-based agricultural communities to increasingly urbanized and industrialized economies (Misselhorn et al., 2012). Prior to the modern era, food production was primarily localized, with people relying on nearby natural resources and small-scale farming to meet their nutritional needs (Boliko, 2019). The influence of food on shaping human societies is evident from the study of human civilization. Wolf suggested, in the Paleolithic era family was considered as a society (Wolf, 1983). Analyzing the standard of living of the people in the

[3] Simon, George-André. History of Food Security [EB/OL] http://www.fao.org/fileadmin/templates/ERP/uni/F4D.pdf. 2012/2020-10-16.

era, Sahlins claimed that the people lived their lives as affluent (Sahlins & Graeber, 2017). Justifying that people living in the Paleolithic era were affluent, Sahlins further mentioned, the people were affluent as their necessities could be found around them (Sahlins & Graeber, 2017). However, over the years people's way of living evolved, so as, the society. If the evolution of the society is examined with people's living standard from economic aspects, one can easily conclude, it is trade, which has played an essential role in the evolution of the modern societies. Mostly, because people in the modern society believe in trade globalization, which urges economic stability, and advance technology. Arancha González Laya, Executive Director of the International Trade Centre, holding the view "trade can help us create a better society" said, "trade gives people a tangible interest in each other's economic well-being and offers a path to fiscally responsible growth."[4] However, if the trade is critically examined with a holistic approach, then one will conclude that food is the core element behind the evolution of modern societies. According to Morillo et al., in the Paleolithic era food was consumed by hunting, however, a relative change in food consumption was brought after the invention of agriculture in the Neolithic era (Morillo et al., 2014). Hueston and McLeod discussed how food system and food trade emerged with agriculture, which brought the socio-economic changes in the societies (Hueston & McLeod, 2012). Over time, food system and food trade got significant, mostly, because of the increase in population growth that had an effect on food availability or security. The nexus between population growth and food availability is still a significant issue; addressed by the Food and Agricultural Organization (FAO) of the United Nations and by World Economic Forum (WEF) in *Global Risk Report 2020.* As addressed by Spitz food security has a long history (Spitz, 1985). The only solution to this enduring issue was and is the trade of surplus food and increase food production. Thence, it won't be imprecise to assert that if food is a problem, then change is perspective can make food itself the solution.

Food trade has opened new doors of development and opportunities in the societies, as people got more involved in farming, manufacturing, and exporting/importing. Wang mentioned Baron von Richthofen's in his book titled *"Ueber die centralasiatischen Seidenstrassen bis zum 2. Jh. n. Chr"* has discovered different trade routes in early centuries (Wang, 2018). These trade routes were also discussed by Liu and Alfed (Andrea, 2014; Belanger, 2011). Studying these trade routes, like Silk Road, Spice road, Waterbourne route, Caravan route, it is observed that all these routes, mostly, carried food since the ancient times. This implies that the food of one society, if we consider society as a country, travels across the borders, and creates an impact on the appetite of people of another society. How this impact on the appetite influenced in developing different cuisines in the society can be apprehended by studying Ken Albala, Joyce E. Haplin, and Paul H Freedman's join work *Food in Time and Place: the American Historical Association Companion to Food History*

[4] Laya, Arancha González, Trade can help us create a better society for all—if we use it properly [EB/OL] https://www.weforum.org/agenda/2016/11/trade-can-help-us-create-a-better-society-for-all/ 2016-11-03/2020-05-21.

(Freedman et al., 2014). This book discusses how American food culture and cuisine developed from ancient Mediterranean and medieval societies.

Other than influencing development in appetite food trade has also influenced agricultural revolution in societies, which is well explored by Paarlberg D. and Paarlberg P. in the book *The Agricultural Revolution of the twentieth Century* (Paarlberg and Paarlberg 2000). Technological and biotechnological revolutions are influenced by food trade, which had an impact on the food chain production, packaging to supply. For instance, food preservative technology this conventional term is said to emerge in the early nineteenth century and is used to inhabit food from spoilage by the growth of bacteria. From the book *Food Preservatives* edited by Russell, and Gould a brief history of food preservatives technology can be perceived which suggests fermentation, oil packing, pickling, salting, and smoking are all ancient preservation technologies (Gould & Russell, 2003). However, over time food preservation technology evolved and at present as organic acids, esters, nitrates, and so forth are used which increased public concern because of food safety. As a consequent of this concern, many scientific innovations emerged to test and ensure that the food traded is safe for the people.

Thence, it is apprehended that food, in terms of availability, is a problem in almost every society over centuries. To resolve it, societies were influenced to trade food, which addressed the problem and instigated economic stability. It is the demand of food that instigated the innovation of technologies, which created an impact on the food chain, from production, processed, to supply and consumption.

1.3.2 Globalization's Impact on Food Security

Food is the paramount necessity of the people and food security is one of the securities that can affect global peace and security. To a certain extent, it can be said that the prosperity and stability of a country, society, and the world all depend on the availability, access, and safety of food. Hence, food security plays a vital role. The global food system has undergone a major transformation in the past century, transitioning from primarily local and regional food production and distribution to a highly globalized and interconnected system (Fears & Canales, 2023). The advent of new technologies, such as refrigeration, transportation, and communication, has enabled the expansion of international food trade, allowing countries to source food from distant regions (Fears & Canales, 2023). Concurrently, the rise of large-scale industrial agriculture, the consolidation of food processing and retailing, and the liberalization of global trade have reshaped the global food landscape (Clapp et al., 2022).

While this globalization of the food system has brought about increased food availability and affordability for many, it has also introduced new vulnerabilities and inequalities. The uneven distribution of resources, power, and influence within the global food system has often led to the marginalization of small-scale producers, particularly in developing countries, and the concentration of control in the hands of a few multinational corporations. These dynamics have significant implications for food security, as they can limit access to food, particularly for the world's most vulnerable populations (Krivonos et al., 2015; Premanandh, 2011).

Additionally, the globalization of the food system has heightened the interconnectedness of food production and distribution, making it more susceptible to shocks and disruptions. So, international food trade also plays an essential role in maintaining food security as it ensures availability of food in the global food system. However, in the modern world, many people are still facing the threat of starvation, which has a serious impact on maintaining social order and stability. In particular, the sudden outbreak of Covid-19 has increased the proportion of hungry people. According to World Food Program (WFP) Reports "at the beginning of 2020, 135 million people around the world were already facing extreme hunger and the figure could rise to a staggering 265 million people by the end of this year." The latest edition of the *State of Food Security and Nutrition in the World* (SOFI) (FAO et al., 2021) stated that "nearly one-tenth of the world population – up to 820 million people went hungry in 2020."[5] The complexities in the modern food system and industry have motivated many countries in recent years to strengthen their legal regimes over food safety and security—that led to various questions. To what extend the national enactments incorporated the recommendations in international food security guidelines? Do the international food security guidelines and law address the impact of climate change on food security? How much effective the contemporary international food guidelines and agreements are in creating cooperation? What are the scope and loopholes of these laws? Is it possible to guarantee food security with international cooperation? If yes, then how to guarantee food security with international cooperation? Whether there is right to food and whether that right imposes an obligation on global community. Is there any food waste law? When the modern food system was struggling to explore the answer to these questions, Covid-19 crisis has introduced new challenges and as the crisis continues to unfold it can be speculated that more challenges are waiting to come. So how is Covid-19 crisis affecting food security more?

UN World Food Programme Executive Director David Beasley said, "In a worst-case scenario, we could be looking at famine in about three dozen countries."[6] Although no major food shortages have emerged as yet, however, as of April 24, 14 countries had food export bans in place for 20 different products (CSIS.org, n.d.). However, most restrictions have been lifted, and by early July, only two countries continued to impose these measures and the situation can again deteriorate (Laborde et al., 2020a) as such food export restrictions was experienced in the 2008–2010 during the global financial crisis. The food trade restrictions do not only affect the global food industry but also affects the four pillars of food security: availability, access, utilization, and stability (Laborde et al., 2020a, b). This clearly

[5] FAO, IFAD, UNICEF, WFP and WHO. The State of Food Security and Nutrition in the World 2021 [R] Rome: FAO, 2021.

[6] WFP.ORG. WFP Chief warns of hunger pandemic as COVID-19 spreads (Statement to UN Security Council) [N/OL] https://www.wfp.org/news/wfp-chief-warns-hunger-pandemic-covid-19-spreads-statement-un-security-council 2020/2020-08-10.

portrays how the contemporary international food guidelines and law have failed to address food governance which requires cooperation and collaboration. Recently, with the initiative of WTO in the 12th Ministerial Conference the members agreed "not to impose export prohibitions or restrictions on foodstuffs purchased for non-commercial humanitarian purposes."[7] After the 2007–2009 food insecurity crises policy partnership on food security is been seen such as, Niigata Declaration on APEC Food Security, 2010, Kazan Declaration, 2012, Beijing Declaration, 2014, Piura Declaration, 2016, ASEAN Cooperation in Food, Agriculture and Forestry (2016–2025), India-Africa Mutual Cooperation in Agriculture, Puerto Varas Declaration on Food Security (2019) and so forth. However, these bilateral, regional, and multilateral cooperations have fallen short to build global food governance that is essential for a sustainable global food system. Mostly, because these laws are trade oriented, do not address the four pillars of food security adequately, lacks harmonization in perspectives and lacks adequate legal binding affect. Here the emerging question is—how to build effective cooperation and collaboration?

Our persisting struggle with food insecurity, even after 100 years, indicates that we have failed to address the root of the food problem causing food insecurity. The imperative problem is in the approach, the notion food insecurity is seen in terms of hunger, poverty, and economic development. However, its multidisciplinary aspect is not emphasized. As food insecurity is understood more in economic terms, hence, there is an unequal distribution of food. Consequently, where there is sufficient food available, accessible, and affordable, food is wasted there. Hence, according to reports the developed countries are the one to waste food the most. On the other side, because of the trade policy framework many developing and least developing countries also waste a large percentage of food in production process. For instance, because of the law of European supermarkets and their direct suppliers, Kenyan smallholders had to waste around 40% of what they grow for European markets—despite the fact that the country holds millions of hungry people (Stuart, 2015). Beside food production stage, food waste is also witnessed through the food supply chain even at the consumption level, i.e. household, restaurants, and canteens and so forth. This mismatch in the understanding of food security is needed to be addressed as it is the key to the long-lasting solution.

Therefore, the legal and policy framework on food security should be built considering the multidisciplinary aspects of food security, such as availability, accessibility, affordability, utilization, and sustainability. Availability and accessibility should not just be seen in economic terms, but in terms of equitable distribution, physical infrastructure, energy, water, climate, environment, education, social inclusion, and good governance (Singer, 1986). Efforts should focus on creating sustainable food systems by addressing environmental constraints, improving efficiency of production systems, reducing food loss and waste, and promoting nutrition-sensitive agriculture.

[7] WTO. Ministerial decision on world food programme food purchases exemption from export prohibitions or restrictions WT/MIN(22)/29 WT/L/1140 [C] 2022 Ministerial Conference Twelfth Session Geneva, 12–15 June:1.

1.3.3 Food System and Food Security: The Role of Trade Policies, Unequal Distribution of Resources and Markets

Food security is not just about having enough food or its availability, but also the ability of people to access food, in economic terms as well as physical terms. Food systems encompass the entire range of activities involved in the production, processing, distribution, preparation, and consumption of food, and the outputs of these activities, including socio-economic and environmental outcomes (Krivonos et al., 2015; Pinstrup-Andersen & Pandya-Lorch, 1998). Food security is therefore inextricably linked to and dependent on the functioning of these food systems (Krivonos et al., 2015). Food systems are shaped by food trade policies and the global food trade. Trade policies can enhance or constrain food security through their impacts on food availability, access, utilization, and stability (Krivonos et al., 2015; Mc Carthy et al., 2018). The contemporary international food instruments, as evident from reviewing Shaw, D. John's book *World Food Security: A history since 1945*, are drafted with the objective of emphasizing food security (Shaw, 2007). Sumner in *Agricultural trade policy and food security* (Sumner, 2000) has clearly mentioned that "food security is a stated objective of agricultural commodity policy in virtually all countries. Rich and poor, importer or exporter, all governments seem to state food security of their population as an objective for agricultural policy." So, it won't be imprecise to argue that the contemporary international food instruments are drafted with the objective of food security. However, if "food insecurity" was focused more than, concentration to address the problem would get divided. For instance, lack of food is one of the rationales of food insecurity. Addressing the problem lack of food, it is observed, the main causes behind it is lack of agricultural land, climate change, need of technology and innovations, increasing population, need of capacity building, need of investment, and unequal distribution of food. However, these factors are not discussed while addressing international and national food instruments.

Considering Steier and Patel book *International Food Law and Policy* known as the first interdisciplinary piece of academic literature of its kind with a comprehensive approach to teaching the major aspects of food regulation, law, policy, food safety, and environmental sustainability in a global context, it is observed that the authors focused on public health, international food trade, the right to food, intellectual property, and global regulatory aspects of food production (Steier & Patel, 2017). However, didn't consider examining the level of cooperation, bilateral and multilateral agreements, and many other international legal instruments that have an impact on the food system. However, the literature gives an in-depth knowledge about the food law and food system of United States and European Union, which will contribute largely in the future research. But the literature focuses more on public health issues than analyzing the laws and policy.

Similar Anderson in *Agricultural Trade, Policy Reforms, and Global Food Security,* noted, food security requires improving the three inter-related elements of food availability, access, and utilization, as well as reducing market instability (K.

Anderson, 2016). The author suggested "openness [...] contributes to three components of food security: availability, access, and market stability." From Anderson's work market's function to the food security and safety is visible. Although the author attempted to establish the connection between availability, access, and market stability, however, the difference between availability and access and its individual connection with market stability remained unclear. According to Rivera food availability is achieved when sufficient quantities of food are consistently available to all individuals within a country, through household production, other domestic outputs, commercial imports or food assistance (W. M. Rivera & Qamar, 2003). The author further defined "food access" as "food access is ensured when households and all individuals within them have adequate resources to obtain appropriate food for a nutritional diet. Access depends upon income available to the household, on the distribution of income within the household and on the price of food."[8] So this indicates, even if sufficient quantities of food are consistently available, market instability can cause food insecurity; where, if the price of food is higher than the distribution of income within the household then food can be available but not accessible. Furthermore, despite the identification of the unequal distribution of food to regions, countries, households, and individuals, the connection between it with market stability is not addressed by the author. Moreover, it can be deduced that for financially challenged people food can be difficult to access. So, food insecurity can occur due to market instability.

On the other hand, scholars like Shaw, Caraher, and Furey in their work have discussed food aid and assistance which is seen from emergency perspective. Shaw has examined the role of UN World Food Programme in food aid and food assistance and relevant rules in this regard. Whereas, Caraher and Furey has scurried the emergency food aid provisions. However, again unequal distribution of food remained undiscovered (Caraher & Furey, 2018; Shaw, 2011). Increasing population with increasing globalization has made food security a global issue. By 2050 it is predicted that the world pollution will reach to 9.8 billion nearly, which means more than 3 billion more mouths to feed.[9]

Whereas, 1.3 billion tons of food are being wasted each year[10] and when 690 million people are not getting adequate health food, number of which might exceed 840 million by 2030.[11] This clearly indicates the unequal distribution of food. Although, unequal distribution of food requires cooperation and collaboration among the countries to ensure that the "food distribution cycle" operates uninterrupted.

---

[8] Rivera, William McLeod et al. Agricultural Extension, Rural Development and the Food Security Challenge [R] Rome: FAO, 2003.

[9] Ranganathanet.J et al. How to Sustainably Feed 10 Billion People by 2050, in 21 Charts [EB/OL] https://www.wri.org/insights/how-sustainably-feed-10-billion-people-2050-21-charts 2018/2020-05-21.

[10] Chainey, Ross. Which countries waste the most food? [EB/OL] https://www.weforum.org/agenda/2015/08/which-countries-waste-the-most-food/ 2015/2020-05-21.

[11] FAO, IFAD, UNICEF, WFP and WHO. The State of Food Security and Nutrition in the World 2020: Transforming food systems for affordable healthy diets [R] Rome: FAO, 2020.

So, actions at national and international levels are required to facilitate the "food distribution cycle."

Furthermore, study on TBT and SPS agreement has also gained attention. Focusing on transparency in TBT and SPS Agreement, Katunana in *Transparency in the WTO SPS and TBT agreements: the Real Jewel in the Crown* explored the provisions of the two WTO agreements, TBT and SPS Agreement, with the aim of purifying domestic trade regulations (Bohnenberger, 2021). Katunana argued for implications of the principles of transparency in TBT and SPS Agreement. Because as a result of the negative integration process followed by WTO agreements, different regulations apply to all WTO members and have the potential to affect international trade. Bohnengerger found that Katunana argued to introduce a framework and practices of transparency principles under TBT and SPS Agreement, and to determine the role of the principles of transparency in the legal and institutional systems of WTO related to dispute settlement (Bohnenberger, 2021).

1.4 Drivers of Food Security and Insecurity

Food security and insecurity are global concerns that have been widely researched in recent years. A range of factors influence food security and insecurity, and understanding these factors is crucial for policymakers to address these challenges effectively. Some of the key factors that have been identified in previous studies as influencing food security and insecurity. Economic factors play a significant role in determining food security. In low-income households and communities with limited access to markets and economic opportunities, food insecurity is more prevalent (Grainger, 2010; McGuire, 2015). Market prices for food can also affect food security, with higher prices often leading to greater food insecurity (K. Anderson et al., 2014; Gregory & Coleman-Jensen, 2013; Hoddinott, 2001; Zhang et al., 2013).

Demographic factors, such as population growth, urbanization, and aging populations, also play a role in food security. Urbanization can lead to a shift away from traditional agriculture and a loss of local food production, while aging populations may require specific types of food and nutritional needs (Bongaarts, 2021). Conflict and political instability are also factors influencing food security. Conflicts can disrupt food supply chains, lead to displacement of populations, and restrict access to food, water, and other resources essential for food security. Political instability can also lead to reduced investment in agriculture, reduced access to markets, and increased food prices, which can further exacerbate food insecurity (Hoddinott, 2001).

Climate change is another significant factor influencing food security. Changes in rainfall patterns, temperature, and extreme weather events can lead to crop failures, reduced yields, and lower food availability, particularly in regions that rely on rainfed agriculture (Bongaarts, 2021; Booth & Saunders, 1985). Climate change also affects food safety by altering the spread of diseases and pests, which can lead to food shortages (Gerber 2013; Tubiello, 2013; Murdiyarso & Kauffman, 2011; W. Chandler et al., 2002; Medina Hidalgo et al., 2021; Wheeler & von Braun, 2013). The global food system and climate change resembles a "tit for tat" relation. Food industry is depended on climate, and the change in climate is the result of greenhouse gas (GHG) emission. Any change in climate directly affects the production of food and its industry. The global community has already confronted the effects of climate change, in terms of natural disasters. Natural disasters are such as flood, hurricanes, droughts, heat waves, storms, and so forth have a strong connection with climate change. Moreover, all these natural disasters affect agriculture, mainly in terms of food production, as agriculture is the core source of food. Therefore, the global food system is affixed with climate and any changes in climate have an impact on the overall food system, from the production to availability. To apprehend the relation between the present food regime and climate change, the challenges confronted by the farmers due to climate change are discussed in brief.

The first challenge confronted by the farmers is the change in precipitation, especially rain and temperature, pattern. Rainfall is the natural source of water for crops, and it keeps the soil moist. Too little rainfall can cause dryness and droughts, as the temperature might get higher than normal. Consequently, the soil gets dry and the production of crops gets affected. Conversely, too much rainfall also effects the production of crops. As too much rainfall can cause floods in the fields, wash away seeds, encourage bacteria and fungus growth, and so forth. All these consequences affect the production and can damage crops. Second challenge for the farmer is water scarcity. Over time the water level in the Earth is decreasing, many countries, including developed, developing, and least developed countries, have confronted water scarcity. According to United Nations Development Programme (UNDP) around 1.2 billion people, or almost one-fifth of the world's population, live in areas of physical scarcity, and 500 million people are approaching this situation (Development Programme United Nations (UNDP), 2006). So, water scarcity means increase in dependence on rainwater. Third challenge is land degradation. Bado and Bationo in their work *Integrated Management of Soil Fertility and Land Resources in Sub-Saharan Africa: Involving Local Communities* addressed climate change as one of the factors of land degradation as climate change is likely to affect soil erosion (Bado & Bationo, 2018). Taking the similar view Ahmad in *Biosurfactants for Sustainable Soil Management* addressed the fourth challenge confronted by the farmers that the change is lifecycle of bud bursts, inserts, pollination, and reproduction due to climate change (Ahmad et al., 2018). Researchers and scientists have blamed unusual climate change for the present locust infestation in Africa and predicted how locust infestation will affect food industry leading to food insecurity.

Fifth challenge is natural disasters, like flood and storm, often termed as act of god. These unpredicted and uncontrollable natural disasters are the result of global warming, for which farmers and food industry suffers. These challenges resulting from climate change constraints the production of food and quality. The WEF in its *Global Risk Report 2020* has predicted food insecurity in coming years and addressed climate change and biodiversity as the rationales for it (WEF, 2020). Therefore, it can be asserted that the stability of the food regime depends on climate change. Tim Wheeler and Joachim von Braun in their work *Climate Change Impacts on Global Food Security* (Wheeler & von Braun, 2013) stated, "climate variability and change will exacerbate food insecurity in areas currently vulnerable to hunger and under nutrition." Furthermore, they have discussed the importance of building "climate-smart food system."[12]

Gerber suggested reduction of emissions may be achieved by reducing production and consumption, by lowering emission intensity of production, or by a combination of the two (P. J. Gerber, 2013). In this regard technologies and practices can play a significant role in reducing emissions and 30% reduction of GHG emissions is possible if producers use technologies and practices with lower emission intensity. However, Glatzle assessing Gerber's work with FAO rebut highlights that the publication ignored the uncertainties associated with the climate sensitivity of GHGs (Glatzle, 2014). According to Cosme improving human diet and promoting reduction in meat and "empty calorie" consumption will contribute in reduction of GHG emission.[13] However, solely focusing on our present diet, ways of consumption and using green technology can't reduce GHG emission, building waste management system is also essential. Therefore, in the context of climate change, the contemporary global food system requires reform in terms of use of green technology, land use, waste disposal mechanism, and way of consumption.

Gerber also suggested practices and technologies that reduce emissions can often simultaneously increase productivity, thereby contributing to food security and economic development (P. J. Gerber, 2013). However, there is no harmonization in use of technologies among the producers, hence, intensities of emission vary greatly between farms to farms, production units, and even within similar production systems. Furthermore, as there is no code of practice for different stakeholders in the global food system, dissimilarity in ways of consumption and waste management is also evident. To bring harmonization in use of technologies and code of practice, Newell, Taylor, and Touni suggested theorizing the Governance of Food and Agriculture, internationally, regionally, nationally, and sub-nationally through policy framework and strategies, for instance, Climate-Smart Agriculture (CSA) in Kenya (Newell et al., 2018). Their suggestion of theorizing the Governance of Food and Agriculture suggests the contemporary global food system does not only lacks in harmonization, code of practice, and use of green technologies, it also lacks food governance and green institutional framework.

Newell, Taylor, and Touni found CSA strategies in practice can be characterized by multi-scalar power dynamics and tensions as policy is translated from the international and regional to the national and sub-national levels (Newell et al., 2018). CSA is new strategy launched with three different approaches—firstly, assessment for the need of the CSA, secondly, use of technologies to improve the efficiency and capacity of farm production, and thirdly, creating a legislative framework. In

[12] The term is used to define an Information and Communications Technology (ICT) and Artificial intelligence (AI) approach to food production that can improve productivity, increase resilience to climate change and reduce greenhouse gas emissions.

[13] Cosme, R. Reducing Food-Related Greenhouse Gas Emissions Through Modification of Human Diet [EB/OL] http://climate.org/reducing-food-related-greenhouse-gas-emissions-through-modification-of-human-diet/ 2017/2020-05-21.

Kenya, CSA strategic plan was launched for a term 2017 to 2026 with six principles—multidimensional, considering short, medium, and long term; County and context specific; intergovernmental and participatory; Sustainable Agriculture; Good governance and Environmental management. CSA is still under process in Kenya, however, its success and failure is yet to be examined.

Tubiello and Medina et al. suggested the contemporary global food system requires development policies in the area of safeguarding food security and rural livelihoods, while implementing adaptation and mitigation measures for climate change (Medina Hidalgo et al., 2021; Tubiello, 2013). He further suggested policies for three different terms—short term, medium term, and long term. However, the effectiveness and enforcement of these policies at national and international levels requires cooperation, coordination, and monitoring in national and international levels to create practical actions. This returns us back to Newell, Taylor, and Touni's concept of *governing of food and agriculture* and points on the importance of governance to develop a framework through adaptation and mitigation policy, and bring harmonization in use of technologies and code of practice (Newell et al., 2018). Other factors like environmental factors such as soil degradation, water scarcity, and loss of biodiversity also impact food production and food security. Degradation of natural resources can lead to reduced yields and lower quality of crops, which can exacerbate food insecurity.

Food waste is another significant factor contributing to food insecurity. Approximately one-third of all food produced is wasted globally, which could otherwise be utilized to feed those who are food insecure (Lipinski et al., 2013; Sinkko et al., 2019; van Geffen et al., 2019). Reducing food waste is an essential step toward addressing food insecurity.

Addressing the various factors influencing food security and insecurity is necessary to ensure that everyone has access to sufficient, safe, and nutritious food. So, a multifaceted approach, which takes into account economic, climate, political, demographic, food waste, and environmental factors, is required to tackle this challenge effectively.

1.5 The Food Waste Trilemma: Climate Change, Food Security, and Mitigation Strategies

Study on food waste is growing; however, the contemporary study has shown continuous efforts on quantifying food waste, reducing food waste by behavioral change, and assessing and reforming food waste management. However, there are still major gaps exists. Such as, existing studies focused on quantifying food waste, mostly, focused on household food waste—in contrast food waste at food service and retail is didn't gain much attention. On the other hand, studies on developing countries like Africa, China, India, and Philippines have gained momentum. Furthermore, many researchers have studied food waste from climate change perspective, and there is a major gap in examining food waste as triptych term by analyzing the relation between food waste, climate change, and food security.

Considering the serious challenges that food waste possesses to the overall sustainability of food production and consumption many researchers have focused on analyzing the relations between the "food waste" and "climate change." Sinkko et al., asserted that the food supply chain has significant economic, social, and environmental impacts (Sinkko et al., 2019). Further acknowledging Godfray, and Beddington, et al., contention (Godfray et al., 2010), the authors stated, "food waste exacerbates food insecurity and malnutrition, and increases pressures on climate, water, and land resources contributing to natural resources depletion and environmental pollution."[14] The study focused on European citizens and aimed at firstly, to calculate the environmental impacts of food consumption and food waste generation along the food chain; secondly, to assess the environmental impacts of food waste, taking into account impacts from waste treatment of wasted food and additional food production, when part of the food is wasted; and lastly, to assess the impact of food waste reduction and diet change when compared to the overall environmental impacts of food consumption. The results of the study showed that food production and food waste holds environmental impacts such as emission. The study further asserted that "agri-food, animal-based food and frozen food hold the highest contribution in emission which can be reduced if food waste at consumption level could be avoided, ranging between 15% and 21% of the total impact of food consumption."[15]

It can be argued that addressing food waste management is imperative to prescribe a legal framework to develop a proactive constructed food waste measurement system that can reduce food waste at all stages. Many of the prescribed remedies for reducing food waste in the contemporary literature have focused on the change in attitude of stakeholders involved in each stage of the global food supply chain. However, these remedies mainly target reducing household food waste. Stefan, et al. conducting survey of 244 Romanian consumers examined the influence of intentions not to waste food, planning, and shopping routines (Stefan et al., 2013). The results of this exploratory study revealed that "planning and shopping routines are determined by moral attitudes towards food waste and perceived behavioral control." The study argued to reduce household food waste, consumers' food waste behavior needs to change and shift toward tools to deal with food waste activities. In a similar study conducted by Jörissen et al. focusing on Italy and German consumers, also argued for instruments to reduce food waste and to increase consumers' awareness (Jörissen et al., 2015).

---

[14] Sinkko, T., Caldeira, C., Corrado, S., & Sala, S. Food consumption and wasted food [J] Saving Food, 2019: pp. 315–346.

[15] Ibid.

Many researchers focusing on reducing household food waste such as, Segrè et al.[16] Quested, et al.,[17] Bernstad (2014), Stancu et al. (2016), Jagau & Vyrastekova (2017), and Principato (2018) have argued for the similar mitigation strategy and creating awareness. However, van Geffen et al. have taken a different approach studying the distinction between interventions that steer toward setting an intention to lower food waste levels and interventions that steer toward implementing such an intention (van Geffen et al., 2019). But very few study focused on building a tool or instruments to reduce food waste that is often argued by researchers a mitigation strategy. For instance, Singh discussed various preventive measures, and how various technology that can be used for the treatment of food waste (Singh, 2020). The finding of the study suggested that to reduce and prevent food waste it is important to use sustainable food waste management approaches. Whereas, Costello studied application of "the concept of Zero Waste" in food waste management which refers to "designing and managing products and processes to systematically avoid and eliminate the volume and toxicity of waste and materials, conserve and recover all resources, and not burn or bury them."[18] Many researchers have studied food waste disposal system from contrasting perspective, but not focused on building anti-food waste system through legal framework. However, many have studied different anti-food waste policies undertaken by governments, but these studies focused on assessing the policies in term of initiatives. Moreover, the national anti-food waste laws are not examined, nor any legal framework, preventive measures, guideline, or initiative has been taken into consideration for study to develop an anti-food waste agreement; except addressing the issue of food waste.

1.6 The Role of Law in Addressing Food Security

The role of law in addressing food security is multifaceted and spans various levels, from international to national. At the international level, human rights law recognizes the right to adequate food as a fundamental human right, placing obligations on states to respect, protect, and fulfill this right. International trade law has significant implications for food systems and access to resources. At national level, agricultural policies, and food laws have shaped food production and distribution. In this section we discuss about international agreements and national-level laws and policies that shape food security.

[16] Segrè, A., & Falasconi, L. Il Libro Nero Dello Spreco Alimentare In Italia.[M] Milano: Edizioni Ambiente, 2011.

[17] Quested, T. E., Marsh, E., Stunell, D., & Parry, A. D. Spaghetti soup: The complex world of food waste behaviours. [J] Resources, Conservation and Recycling, 2013 (79): 43–51.

[18] Costello, C. The concept of Zero Waste [J] Saving Food, 2019 pp. 369–391.

1.6.1 *Understanding Laws Addressing Food Security at International Level*

The right to food is a fundamental human right recognized under international human rights. A human rights perspective on food security necessitates a shift from viewing food solely as a commodity to recognizing it as a fundamental human right. This perspective has profound implications for international food law and the pursuit of global food security. The right to food, enshrined in international human rights instruments ("Food Sovereignty, Food Security and the Right to Food" 2021; Ziegler et al., 2011a) imposes obligations on states to ensure access to adequate food for all, particularly for vulnerable populations. However, the current international legal framework, primarily focused on facilitating trade and harmonizing food safety standards, falls short of effectively addressing this right.

A critical analysis reveals a gap between the rhetoric of the right to food and its practical implementation. While numerous international declarations and agreements acknowledge these right concrete mechanisms for enforcement and accountability remain weak. This is evident in the continued prevalence of hunger and malnutrition globally, despite the existence of a legal framework supposedly guaranteeing this right. McClain-Nhlapo (2004) and Ziegler et al., (2011b) discuss indicators of food security, highlighting the complexities of measuring and addressing this issue.

Furthermore, the current international food law framework often fails to address the structural inequalities that perpetuate food insecurity and favor trade liberalization and market-based solutions. This neglects the root causes of food insecurity and hinders efforts to achieve sustainable and equitable food systems. This demonstrates the lack of attention given to these crucial issues in existing international food instruments. The Covid-19 pandemic has further exposed the fragility of the global food system and the inadequacy of current legal frameworks. The pandemic exacerbated existing inequalities and highlighted the need for stronger international cooperation and a human rights-based approach to food security. Laborde et al. (2020a, b), and Smith & Wesselbaum (2020) emphasized on the importance of a human rights approach in addressing the health implications of food insecurity, particularly in the context of global challenges like climate change and pandemics.

To effectively address the right to food, international food law must move beyond a narrow focus on trade and safety standards. It must incorporate a human rights framework that addresses the root causes of food insecurity, promotes equitable access to resources, and ensures accountability for states' obligations. This requires a shift in focus from market-driven solutions to a human-centered approach that prioritizes the well-being of all people. The proposed International Food Security Treaty of 1993, though unsuccessful, offers valuable insights into the potential for a more comprehensive and rights-based approach to global food security. By incorporating the principles of the right to food into international law and practice, we can move toward a more just and sustainable global food system.

Besides, WTO agreements, TBT and SPS Agreement, also focused on right to food, food sovereignty, and so on. *The Fight for the Right to Food: Lessons Learned,* a work by Ziegler et al., added a different view to the concept food security by defining the right to food and food security and demonstrating that to understand food security, it is necessary to understand the right to food (Ziegler et al., 2011a). Holding the similar view Schanbacher examined the right to food from the perspective of international law and national security is addressed and discuss (Schanbacher, 2010).

Furthermore, giving importance to food governance, McKeon in *Global Food Security Governance* explored the major dilemma of global food security governance from the institutional dimension and how the food security regulation evolves into the global food security system with the help of the institutions (McKeon, 2021). He argues, the problematic interaction between the constituent systems in the global food security system stems from the different perceptions of what food means by different actors, such as food as economic commodities, food as ecosystem services, and food as basic needs or human rights. The dilemma of global food security governance lies not only in the pluralism and conflict of food norms, but also in the imbalance between them. It is further argued that the dilemma represented by the complex system of global food security is not only the result of power game between developed and developing countries.

The previous literature on the food security was limited to analyzing the contemporary agreements and law, such agriculture law, TBT and SBP agreements, or agues for right to food. But previous literatures do not analyze to what extend these contemporary international food guidelines has succeeded to address food security from food safety angle. However, food security is not analyzed from food insecurity angle, nor a multidisciplinary approach is adopted. Despite acknowledging the need of cooperation and collaboration in the contemporary global food system, the literates do not provide a guideline. Some has addressed the role of international organization in food security and food safety from the perspective of multilateral governance framework for food safety. Such as Alemanno in *The Multilateral Governance Framework for Food Safety: A Critical and Normative Overview,* has given an overview of international organizations mandated to work for food security and safety, such as, Food and Agriculture Organization (FAO), the World Health Organization (WHO), Codex Alimentarius Commission (CAC), the World Animal Health Organization (OIE), the World Trade Organization (WTO), the United Nations Environment Program (UNEP), the Organization for Economic Cooperation and Development (OECD), and International Food Safety Authorities Network (INFOSAN) (Alemanno, 2015). Alemanno found the lack of adequate harmonization and connection between the international organizations dealing with food safety, food security, and trade law are quite visible. Furthermore, asserted that the world currently lacks of a global international food system and the Codex standards referred by the SPS framework establishes food safety standards for the use of its Members, which can be reconstructed to "appropriate level of protection" depending on "the Members assessment of risk (Alemanno, 2015).

1.6.2 Understanding Laws Addressing Food Security at National Level

Most countries have adopted specific laws and regulations to address food security and food safety at the national level. Analyzing U.S. agricultural policy in The *Ethics of Aid and Trade,* Thompson noted, "The primary motive for trade, however, is neither diplomatic nor humanitarian, but profit."[19] Later, the author noted, "trade would then be allocated to the monetary or economic goals of a nation".[20] Hence, the author argued, "the philosophical basis for food policy is obsessed with two others issues except surplus – national interest in food security and conflict between property rights and distributive justice."[21] However, enactments such as US, *Federal Food and Drugs Act,* and *US, Food Quality Protection Act* remained untouched.

Holding a similar view about China, Huang and Rozelle in *Agricultural development and policy before and after China's WTO accession* (Huang & Rozelle, 2008), has investigating the agricultural policy of China, stated, "Food security—one of the issues of central concern to policymakers in China." Huang and Rozelle focused particularly on the agricultural policy of China and the development of the sector after China's WTO accession without focusing on the 10% arable land, capacity building of the farmers, climate change, and so forth. When many researchers focused on agriculture policy of China, Louis Augustin-Jean and Xie in *Food Safety Standards and Market Regulations as elements of Competition—Case Studies from China's International Trade* focused food safety standard discussing how China is influencing the development of global norms in food safety and, beyond that, the international dairy market (Augustin-Jean & Xie, 2016).

Furthermore, focusing on the global pandemic and its impact on food 丁冬 (Ding Dong) in his article《疫情经济与粮食安全治理的"政策之锚》*(The "Policy Anchor" of Epidemic Economy and Food Safety Governance)* focus on the food safety governance from China's perspective and argued how to improve legislation and innovate supervision to achieve refined food safety governance (丁冬, 2020). However, enactments such as Regulations on the Implementation of the Food Safety Law of the People's Republic of China and the Measures for the Administration of New Food Additives remained unexplored.

Alemanno and Gabbi examined the food safety law of EU in *Foundations of EU Food Law and Policy: Ten Years of the European Food Safety Authority* by analyzing European Food Safety Authority (Alemanno & Gabbi, 2014). Kai P. Purnhagen et al. in *Beyond Food Safety—EU Food Information Standards as a Facilitator of Political Consumerism and International Law Enforcement Mechanism* analyzed EU food law and policy, and international law (Purnhagen et al., 2020). For the analysis he studied *the Case C-363/18 Organisation juive européenne, Vignoble Psagot Ltd v Ministrede l'Économie et des Finances ("Occupied Territories case").* However, the

[19] Thompson, Paul B. The Ethics of Aid and Trade: U.S. food policy, foreign competition, and the social contract [M] Australia: Cambridge University Press, 1992: p. 22.

[20] Ibid.

[21] Ibid. at 25.

focus of the analysis was on the evolving role of information in EU consumer and food law, political consumerism in EU internal market law, and consumer choice and the enforcement of international law. Although the author has not explicitly discussed about food labeling but has touched the issue of food labeling and its influence on the consumer to purchase and eating patterns. He further discussed how food labeling in seen in EU food law for consumer protection. It is observed that most of study focused on the agriculture policies and food safety not on the food law nor examined the loopholes in the law, neither examined how much these laws adhere to the international legal food instruments. Furthermore, it does not examine the national food system, not only in economic or food safety terms, nor in food security terms. However, all these contribute in the present and future research to understand the economic perspectives and health of the countries, by giving the study a multidisciplinary approach that food security and food insecurity deserves.

1.7 Global Food Security Governance

Food security governance is critical at both the global and national levels to ensure equitable access to food and prevent food crises. Global food security governance, while framed by international agreements and organizations, faces significant challenges in effectively addressing persistent food insecurity. The existing framework, comprising instruments like agricultural laws, the TBT and SPS Agreements of the WTO, and the right to food discourse, lacks cohesive implementation and often prioritizes economic interests over equitable access to food. While organizations like the FAO, WHO, and Codex Alimentarius contribute to food safety and standard setting their efforts are fragmented, lacking a unified global food system (Ayala & Meier, 2017). So, coordination between different international organizations is often lacking, leading to fragmented and sometimes conflicting approaches. This fragmentation, coupled with the "appropriate level of protection" clause in the SPS Agreement, allows states to prioritize national economic interests, potentially undermining global food security goals. Moreover, securing consistent funding and political commitment from member states can also be difficult. Additionally, power imbalances between developed and developing countries can hinder the equitable distribution of resources and decision-making authority.

At the national level, food security legislation often focuses on domestic agricultural policy and food safety, neglecting broader systemic issues. Studies on US and Chinese agricultural policies reveal a prioritization of national economic interests and trade over equitable food distribution and sustainable practices. While some research touches upon food labeling and consumer protection within the EU food law framework, a comprehensive analysis of how national laws adhere to international food instruments and address food system challenges is lacking. Furthermore, the interplay between national food systems, economic considerations, food safety, and overall food security remains underexplored.

This fragmented governance landscape presents both challenges and opportunities. The lack of a cohesive global food system, coupled with the dominance of national economic interests, hinders effective action on food insecurity. However, the existing framework of international organizations and agreements provides a foundation for building stronger global cooperation. The opportunity lies in shifting from a fragmented, market-driven approach to a more integrated, human rights-based framework that prioritizes equitable access to safe and nutritious food for all. This requires strengthening international cooperation, harmonizing national food laws with international instruments, and promoting greater transparency and accountability in food governance mechanisms. Furthermore, integrating a multidisciplinary approach that considers the interconnectedness of food systems, economic policies, environmental sustainability, and social equity is crucial for achieving meaningful progress toward global food security.

1.8 Conclusion

This chapter identifies that the existing international instruments, agreements, and governance mechanisms related to food security have significant gaps and limitations. The identified major problems are: (1) The approach toward food security; (2) Absence of Comprehensive International Food Law; (3) Absence of an operating Global Food System; (4) Lack of Food Governance; (5) Absence of application of legal principles in the global food system; (6) Unequal distribution of food; (7) Absence of anti-food waste provision; and (8) Lack of Cooperation, Collaboration and Harmonization. The contemporary international food law has formed a fragmented multilateral framework primarily focused on food safety in food trade, which falls short in comprehensively addressing the multifaceted issues of food insecurity and building sustainable food security. An in-depth analyses of the contemporary international food law indicates imperative, but disdain issues such as, lack of harmonization in food safety and food standard guidelines, disparity between developing countries and developed countries, and so forth.

However, the contemporary international food law so far has concentrated primarily on understanding the food security under economic perspective and understanding the disputes from food safety perspective. And many relevant issues like climate change, rights and obligations, extend of cooperation among different stakeholders, disparity between developing countries and developed countries, exploring the issues of consensus, exploring the general principles for food law, and so forth are the areas which remained unexplored. Notwithstanding the contemporary researches on food security contributes largely in establishing how food waste exacerbates food insecurity by quantifying food waste data and in developing a legal mechanism and institutional framework to build sustainable and equitable global food system. But falls-shorts in addressing general principles, rights and obligations, anti-food waste guideline, food management guidelines, food standard and food safety, the amplitude of international cooperation and coordination, the fields and ways to cooperate on food security, the principles, obligations, rights and duties of cooperation among the global community and international organizations, and so forth.

Generally thinking about "food law" it is believed to be a legislation which regulates the production, trade, and handling of food, so, it covers the regulation of food control, food safety, and relevant aspects of food trade. Such as minimum quality requirements to ensure the foods produced are unadulterated and are not subjected to any fraudulent practices intended to deceive the consumer. However, analyzing the 100 years of the contemporary global food system a different scenario is observed. Firstly, the contemporary international food law focuses on harmonizing food standard and food safety measures, in such a way so that food trade is not hampered. Secondly, it is observed that there are multiple international instruments which are neither strictly binding in nature nor completely lack legal significance, such as Codex Alimentarius. The only two international instruments creating legal bindings are SPS Agreement and TBT Agreement. Thirdly, WTO agreements like the SPS Agreement have changed the status and effectiveness of these food safety standards ("Does Science Speak Clearly and Fairly in Trade and Food Safety Disputes? The Search for an Optimal Response of WTO Adjudication to Problematic International Standard-Making" 2013). Fourthly, National legal frameworks play as a key pillar in an effective global food control system. In all countries, food is governed by a complexity set of laws and regulations; hence, there is lack of harmonization and corporation in each and every level of the global food system.

Hence, it can be argued that the contemporary global food system in its current configuration is not economically, ecologically, or socially sustainable and equitable. As the contemporary international instruments regulating the contemporary global food system focuses on food safety standards, not focusing on food security in terms of sufficient quantity of food, and operates an unfair global food system with diverse and conflicting food safety standards which presents the disparity between developed and developing countries.

References

Ahmad Z, Imran M, Qadeer S, Hussain S, Kausar R, Dawson L, Khalid A (2018) Biosurfactants for sustainable soil management. In: Advances in agronomy. Elsevier, pp 81–130. https://doi.org/10.1016/bs.agron.2018.02.002

Alemanno A (2015) The multilateral governance framework for food safety: a critical and normative overview. In: Food safety, market organization, trade and development. Springer International Publishing, pp 9–43. https://doi.org/10.1007/978-3-319-15227-1_2

Alemanno A, Gabbi S (2014) Foundations of EU food law and policy: ten years of the European food safety authority. In: Foundations of EU food law and policy: ten years of the European food safety authority. Routledge, London

Anderson K (2016) Agricultural trade, policy reforms, and global food security. Palgrave Macmillan US. https://doi.org/10.1057/978-1-137-46925-0

Anderson K, Ivanic M, Martin WJ (2014) 8. Food price spikes, price insulation, and poverty. Comment. In: The economics of food price volatility (pp. 311–344). University of Chicago Press. https://doi.org/10.7208/chicago/9780226129082.003.0009

Andrea AJ (2014) The silk road in world history: a review essay. Asian Rev World Hist 2(1). https://doi.org/10.12773/arwh.2014.2.1.105
Ashley JM (2016) Food security in the developing world. Academic Press
Augustin-Jean L, Xie L (2016) Food safety standards and market regulations as elements of competition – case studies from China's international trade. Asian J WTO Int Health Law Policy 11(2). https://doi.org/10.2139/ssrn.2849015
Ayala A, Meier BM (2017) A human rights approach to the health implications of food and nutrition insecurity. Public Health Rev 38(1). https://doi.org/10.1186/s40985-017-0056-5
Bado VB, Bationo A (2018) Integrated management of soil fertility and land resources in sub-Saharan Africa: involving local communities. In: Advances in agronomy. Elsevier, pp 1–33. https://doi.org/10.1016/bs.agron.2018.02.001
Baer-Nawrocka A, Sadowski A (2019) Food security and food self-sufficiency around the world: a typology of countries. PLoS One 14(3):e0213448. https://doi.org/10.1371/journal.pone.0213448
Belanger WA (2011) The silk road in world history. By Xinru Liu. Oxford: Oxford University Press, 2010. x, 168 pp. 74.00 (cloth) 19.95 (paper). J Asian Stud 70(4):1156–1157. https://doi.org/10.1017/s0021911811001975
Bernstad A (2014) Household food waste separation behavior and the importance of convenience. Waste Manag 34(7):1317–1323. https://doi.org/10.1016/j.wasman.2014.03.013
Bohnenberger F (2021) Transparency in the WTO SPS and TBT agreements: the real jewel in the crown by Marianna B. Karttunen Cambridge University Press, 2020. World Trade Rev 20(3). https://doi.org/10.1017/s1474745620000579
Boliko MC (2019) FAO and the situation of food security and nutrition in the world. J Nutr Sci Vitaminol 65. https://doi.org/10.3177/jnsv.65.S4
Bongaarts J (2021) FAO, IFAD, UNICEF, WFP and WHO. The state of food security and nutrition in the world 2020. Transforming food systems for affordable healthy diets FAO, 2020, 320 p. Popul Dev Rev 47(2):558. https://doi.org/10.1111/padr.12418
Booth TH, Saunders JC (1985) Applying the FAO guidelines on land evaluation for forestry. For Ecol Manag 12(2):129–142. https://doi.org/10.1016/0378-1127(85)90080-5
Caraher M, Furey S (2018) The economics of emergency food aid provision. Springer International Publishing. https://doi.org/10.1007/978-3-319-78506-6
Clapp J, Moseley WG, Burlingame B, Termine P (2022) Viewpoint: the case for a six-dimensional food security framework. Food Policy 106. https://doi.org/10.1016/j.foodpol.2021.102164
CSIS.org (n.d.) Covid-19 and food security. Retrieved August 22, 2020, from https://www.csis.org/taxonomy/term/1686
Development Programme United Nations (UNDP) (2006) Beyond scarcity: power, poverty and the global water crisis. UN. https://doi.org/10.18356/334c604b-en
Does science speak clearly and fairly in trade and food safety disputes? The search for an optimal response of WTO adjudication to problematic international standard-making (2013) In: Science and technology in international economic law. Routledge, pp 48–70. https://doi.org/10.4324/9780203381700-11
FAO (2021) The state of food and agriculture 2021. Making agrifood systems more resilient to shocks and stresses. Food and Agriculture Organization of the United Nations
FAO, IFAD, UNICEF, WFP and WHO (2021) The state of food and agriculture 2021. https://doi.org/10.4060/cb4476en
Fears R, Canales C (2023) The role of science, technology and innovation in transforming food systems globally. In: Science and innovations for food systems transformation. https://doi.org/10.1007/978-3-031-15703-5_44
Food sovereignty, food security and the right to food (2021) In: Handbook of critical Agrarian studies. Edward Elgar Publishing. https://doi.org/10.4337/9781788972468.00035
Freedman P, Chaplin JE, Albala K (2014) Food in time and place: the American historical association companion to food history. https://doi.org/10.5860/choice.189462

Gerber PJ (2013) Tackling climate change through livestock – a global assessment of emissions and mitigation opportunities. In: Most, vol 14, no 2. Food and Agriculture Organization of the United Nations (FAO), Rome

Glatzle A (2014) Questioning key conclusions of FAO publications 'livestock's long shadow' (2006) appearing again in 'Tackling climate change through livestock' (2013). Pastoralism 4(1). https://doi.org/10.1186/2041-7136-4-1

Godfray HCJ, Beddington JR, Crute IR, Haddad L, Lawrence D, Muir JF, Pretty J, Robinson S, Thomas SM, Toulmin C (2010) Food security: the challenge of feeding 9 billion people. Science 327(5967):812–818. https://doi.org/10.1126/science.1185383

Gould GW, Russell NJ (2003) Major, new, and emerging food-poisoning and food-spoilage microorganisms. In: Food preservatives. Springer US, pp 1–13. https://doi.org/10.1007/978-0-387-30042-9_1

Grainger M (2010) World summit on food security (UN FAO, Rome, 1618 November 2009). Dev Pract 20(6):740–742. https://doi.org/10.1080/09614524.2010.491540

Gregory CA, Coleman-Jensen A (2013) Do high food prices increase food insecurity in the United States? Appl Econ Perspect Policy 35(4):679–707. https://doi.org/10.1093/aepp/ppt024

Hoddinott J (2001) Choosing outcome indicators of household food security. In: Food security in practice: methods for rural development projects

Huang J, Rozelle S (2008) Agricultural development and policy before and after China's WTO accession. In: Agriculture and food security in China: what effect WTO accession and regional trade arrangements? https://doi.org/10.22459/afsc.06.2008.02

Hueston, W., & McLeod, A. (2012). Overview of the Global Food System: Changes over Time/Space and Lessons for Future Food Safety. In Overview of the Global Food System: Changes over Time/Space and Lessons for Future Food Safety. *In Improving food safety through a one health approach: Workshop summary* (p. 189). National Academies Press.

Jagau HL, Vyrastekova J (2017) Behavioral approach to food waste: an experiment. Br Food J 119(4):882–894. https://doi.org/10.1108/bfj-05-2016-0213

Jörissen J, Priefer C, Bräutigam K-R (2015) Food waste generation at household level: results of a survey among employees of two European research Centers in Italy and Germany. Sustainability 7(3):2695–2715. https://doi.org/10.3390/su7032695

Kang H (2015) A study on the relationship between international trade and food security: evidence from less developed countries (LDCs). Agric Econ (Czech Republic) 61(10). https://doi.org/10.17221/246/2014-AGRICECON

Krivonos E, Morrison J, Canigiani E (2015) Trade and food security: links, processes, and prospects. Front Econ Global 15. https://doi.org/10.1108/S1574-871520150000015002

Laborde D et al (2020a) COVID-19 food trade policy tracker. IFPRI. https://www.ifpri.org/project/covid-19-food-trade-policy-tracker

Laborde D, Martin W, Swinnen J, Vos R (2020b) COVID-19 risks to global food security. Science 369(6503):500–502. https://doi.org/10.1126/science.abc4765

Lipinski B, Hanson C, Lomax J, Kitinoja L, Waite R, Searchinger T (2013) Reducing food loss and waste. Installment 2 of creating a sustainable food future. Washington, DC: World Resource Institute, June

Maxwell S, Smith M (1992) Household food security: a conceptual review. In: Household food security: concepts, indicator, measurement. UNICEF/IFAD, New York/Rome

Mc Carthy U, Uysal I, Badia-Melis R, Mercier S, O'Donnell C, Ktenioudaki A (2018) Global food security – issues, challenges and technological solutions. Trends Food Sci Technol 77. https://doi.org/10.1016/j.tifs.2018.05.002

McClain-Nhlapo C (2004) Implementing a human rights approach to food security, 2020 Africa conference brief, 13

McGuire S (2015) FAO, IFAD, and WFP. The state of food insecurity in the world 2015: meeting the 2015 international hunger targets: taking stock of uneven Progress. Rome: FAO, 2015. Adv Nutr 6(5):623–624. https://doi.org/10.3945/an.115.009936

McKeon N (2021) Global food governance. Development (Basingstoke) 64(1–2):48–55. https://doi.org/10.1057/s41301-021-00299-9

Medina Hidalgo D, Nunn PD, Beazley H (2021) Challenges and opportunities for food systems in a changing climate: a systematic review of climate policy integration. Environ Sci Policy 124. https://doi.org/10.1016/j.envsci.2021.07.017

Misselhorn A, Aggarwal P, Ericksen P, Gregory P, Horn-Phathanothai L, Ingram J, Wiebe K (2012) A vision for attaining food security. In: Current opinion in environmental sustainability, vol 4, no 1. https://doi.org/10.1016/j.cosust.2012.01.008

Morillo S, Black J, Lococo P (2014) War in world history: society, technology, and war from ancient times to the present. Volume 1, to 1500. Psychol Sci 25(9)

Murdiyarso, D., & Kauffman, J. B. (2011). *Addressing climate change adaptation and mitigation in tropical wetland ecosystems of Indonesia* (Vol. 41). Center for Internatinal Forestry Research (CIFOR).

Newell P, Taylor O, Touni C (2018) Governing food and agriculture in a warming world. Global Environ Politics 18(2):53–71. https://doi.org/10.1162/glep_a_00456

Olper A, Curzi D, Bedin E, Swinnen J (2014) Food security, health and trade liberalization. Department of Economics Management and Quantitative Methods

Olper A, Curzi D, Swinnen J (2018) Trade liberalization and child mortality: a synthetic control method. World Dev 110:394–410. https://doi.org/10.1016/j.worlddev.2018.05.034

Paarlberg D, Paarlberg P (2000) The agricultural revolution of the 20th century. In: The agricultural revolution of the 20th century. Lowa State University Press. https://doi.org/10.1002/9780470290170

Pinstrup-Andersen P, Pandya-Lorch R (1998) Food security and sustainable use of natural resources: a 2020 vision. Ecol Econ 26(1). https://doi.org/10.1016/S0921-8009(97)00067-0

Premanandh J (2011) Factors affecting food security and contribution of modern technologies in food sustainability. J Sci Food Agric 91(15). https://doi.org/10.1002/jsfa.4666

Principato L (2018) The complexity of food waste at consumption level: definitions, data, causes and impacts. In: SpringerBriefs in environmental science. Springer International Publishing, pp 1–13. https://doi.org/10.1007/978-3-319-78887-6_1

Purnhagen KP, van Zeben J, Oosterveer P, Ahlborn C (2020) Beyond food safety: EU food information standards as a facilitator of political consumerism and international law enforcement mechanism. Eur L Rev 45(4)

Reutlinger S, Knapp K (1980) Food security in food deficit countries. World Bank Staff working paper, no 393

Rivera WM, Qamar KM (2003) Agricultural extension, rural development and the food security challenge. Food and Agriculture Organization of the United Nations

Sahlins M, Graeber D (2017) Stone age economics. Routledge. https://doi.org/10.4324/9781315184951

Schanbacher WD (2010) The politics of food: the global conflict between food security and food sovereignty. Choice Reviews Online 47(12). https://doi.org/10.5860/choice.47-6944

Shaw DJ (2007) World food security: a history since 1945. Palgrave Macmillan. https://doi.org/10.1057/9780230589780

Shaw DJ (2011) Transforming food aid to food assistance. In: The worlds largest humanitarian agency. Palgrave Macmillan UK, pp 262–314. https://doi.org/10.1057/9780230316713_6

Singer P (1986) Famine, affluence, and morality. In: International ethics. Princeton University Press, pp 247–262. https://doi.org/10.1515/9780691241869-014

Singh K (2020) Sustainable food waste management: a review. In: Sustainable food waste management. https://doi.org/10.1007/978-981-15-8967-6_1

Sinkko T, Caldeira C, Corrado S, Sala S (2019) Food consumption and wasted food. In: Saving food. Elsevier, pp 315–346. https://doi.org/10.1016/b978-0-12-815357-4.00011-0

Smith MD, Wesselbaum D (2020) COVID-19, food insecurity, and migration. J Nutr 150(11):2855–2858. https://doi.org/10.1093/jn/nxaa270

Spitz P (1985) The right to food in historical perspective. Food Policy 10(4):306–316. https://doi.org/10.1016/0306-9192(85)90022-3

Stancu V, Haugaard P, Lähteenmäki L (2016) Determinants of consumer food waste behaviour: two routes to food waste. Appetite 96:7–17. https://doi.org/10.1016/j.appet.2015.08.025

Stefan V, van Herpen E, Tudoran AA, Lähteenmäki L (2013) Avoiding food waste by Romanian consumers: the importance of planning and shopping routines. Food Qual Prefer 28(1):375–381. https://doi.org/10.1016/j.foodqual.2012.11.001

Steier G, Patel KK (eds) (2017) International food law and policy. Springer International Publishing. https://doi.org/10.1007/978-3-319-07542-6

Stuart T (2015) Food: how much does the world need? World Economic Forum (WEF), May 7. https://www.weforum.org/agenda/2015/05/food-how-much-does-the-world-need/

W. Chandler, R. Schaffer, Z. Dadi, P. R. Shukla, F. Tudela, O. Davidson and S. Alpan-Atamar (2002) Climate Change Mitigation in Developing Countries, Report, The Pew Center on Global Climate Change, Washington DC, Bhagirath Jogdand. http://international-environmental-affairs.suite101.com/article.cfm/climate_change_impact_on_food_security/2020-05-21

Sumner DA (2000) Agricultural trade policy and food security. Q J Int Agric 39(4):395–409

Tubiello F (2013) Climate change adaptation and mitigation: challenges and opportunities in the food sector. Natural Resources Management and Environment Department, FAO

Valdes A (1981) Assessing food insecurity based on national aggregates in developing countries. In: Food security for developing countries. Routledge, p 27. https://doi.org/10.4324/9780429048289

van Geffen L, van Herpen E, van Trijp H (2019) Household food Waste – how to avoid it? An integrative review. In: Food Waste Management. Springer International Publishing, pp 27–55. https://doi.org/10.1007/978-3-030-20561-4_2

Wang Y (2018) Belt and road initiative: mutual connectivity of the world. China and the World 1(4). https://doi.org/10.1142/s2591729318500232

WEF (2020) The global risks report 2020 insight report, 15th edn. Weforum.Org

Wheeler T, von Braun J (2013) Climate change impacts on global food security. Science 341(6145):508–513. https://doi.org/10.1126/science.1239402

Wolf ER (1983) Europe and the people without history. Ann Hist Sci Soc 38(6). Cambridge University Press (CUP). https://doi.org/10.1017/s0395264900083335

Zhang Q, Jones S, Ruhm CJ, Andrews M (2013) Higher food prices may threaten food security status among American low-income households with children. J Nutr 143(10):1659–1665. https://doi.org/10.3945/jn.112.170506

Ziegler J, Golay C, Mahon C, Way S (2011a) The fight for the right to food: lessons learned. Choice Reviews Online (07). Springer. https://doi.org/10.5860/choice.49-3926

Ziegler J, Golay C, Mahon C, Way S-A (2011b) The fight for the right to food. Palgrave Macmillan UK. https://doi.org/10.1057/9780230299337

丁冬 (2020) 疫情经济与食品安全治理的"政策之锚. 中国食品药品监管 10:82–89

Chapter 2
Understanding the Relationship Between Food Insecurity, Security, and Safety in the Global Food System

2.1 Introduction

Even after approximately 2.5 million years of human civilization, in this digital globalized Industrial Era our foods are produced through agriculture. With the enhancement of technology, human innovation, and human preferences, the nature of food has transformed from purely agriculture to processed food—the world has witnessed evolution of food. Evolution of food has given a new aspect to food transforming it as commodity and contributing largely in economy. However, with increase in world population and shortage of agricultural lands, many countries have witnessed food insecurity. The issue of food insecurity remained as a camouflaged global issue as it was considered mostly in country or regional levels, however, became perceptible with the global economic crisis in 1929 which had an impact on the then free trade policy and international trade.[1] The global economic crisis and the later events have presented "food insecurity" as a global issue and the urged of global food system. The contemporary global food system is inefficaciously, as it is still evolving with time, technology, human innovations, and human preferences. It took nearly 100 years for the world to understand the importance of food in human life which is beyond being just a biological necessity, as food security is one of the major global securities playing a significant role in human culture, civilization, society, identity, and economy.

Major key phrases that run through this thesis is "food insecurity," "food security," and "food safety." Cognizant of a diverse understanding of the term, this chapter aims to contextualize the discussion by familiarizing the readers with these major

[1] Ruoshi, Hou. History shows trade protectionism is short-lived [EB/OL] http://chinaplus.cri.cn/opinion/opedblog/23/20180822/173553.html. 2018/2020-10-20.

Part of this Chapter got published in Asian Journal of WTO & International Health Law and Policy 2023, 18(2) (SSCI Journal).

S. S. Silvee, *The Law of Food Security: The Legal Framework for a Sustainable and Equitable Global Food System*, LITES - Legal Issues in Transdisciplinary Environmental Studies 5, https://doi.org/10.1007/978-981-95-3636-8_2

key phrases which will contribute in comprehending the major key phrases. It will review the present literature, without offering a detailed discussion of the historical aspect of this issue, rather, focusing on the classifications different national and international organizations, and academic scholarship have produced in relation to the comprehending the concepts. It will examine whether food security and food safety are interchangeable concepts or food safety is an integral part of food security as these conceptions can provide useful guidance to analysis the contemporary international food law and understand whether these contemporary international food law focuses on building food security or attempts to ensure food safety. Comprehending these major key phrases is essential for food policy interventions as it contributes in identifying targeted issues and probable effective ways. This chapter seeks to analyze the first research question: How do food insecurity, food security, and food safety interact and influence each other in the global food system and what are the implications for policy and practice?

This chapter is divided into four segments, including introduction and conclusion. The discussion begins with the attempt to comprehend the concepts "food insecurity," "food security," and "food safety." Later it discusses the disparities, differences, and similarities between the concepts which establish strong nexus. Lastly, the chapter concludes by arguing that "food insecurity" and "food security" is the opposite sides of the same coin, hence, "food insecurity" is the antithesis of the notion "food security," in terms of availability, accessibility, affordability, utilization, and stability, and food safety is an integral part of food security acting as a security system with a set of decisive measures to ensure mutual survival and safety of human health.

2.2 Scrutinizing the Three Major Notions: "Food Insecurity," "Food Security," and "Food Safety"

This section discusses the definitions of the major key phrases "food insecurity," "food security," and "food safety" provided by the national and international organizations which are most dominant in the existing literature and most used in theory and practice in any discussion and commentary on "food insecurity," "food security," and "food safety"; as comprehending these major key phrases contributes in identifying targeted issues and effective ways for food policy interventions.

2.2.1 Food Insecurity

Food insecurity (FIS) is a critical social condition, which is gradually exacerbating, despite progressive economic expansion. Although FIS has a long persisting history, however, the concept was first brought before the international arena in the early 1935 with the report titled as *Nutrition and Public Health* prepared by the Health

Committee of the Health Organization (Russell 1953). However, the committee didn't define the concept with appropriate definition, rather presented the concept as social condition of malnutrition and hunger. Later, in 1974 the World Food Conference FAO presented the concept as relationship between hunger, poverty, and unemployment (Anderson and Cook 1999). FAO defined FIS as a milieu where "a person is food insecure when they lack regular access to enough safe and nutritious food for normal growth and development and an active and healthy life. This may be due to unavailability of food and/or lack of resources to obtain food."[2] Later FAO stated FIS "exists when people do not have adequate physical, social, and economic access to sufficient, safe, and nutritious food, which meets their dietary needs and food preferences for an active and healthy life."[3] Further FAO defined FIS as "a situation that exists when people lack secure access to sufficient amounts of safe and nutritious food for normal growth and development and active and healthy life."[4] So, according to FAO, individuals whose food intake decreases below their minimum calorie requirements, is physically not fit, does not have economic access to sufficient, safe, and nutritious food, have some disease due to the lack of food and hunger or unbalance diet falls under the FIS category (FAO 2000). Similar to FAO's definition, United States Department of Agriculture (USDA) described FIS as "limited or uncertain availability of nutritionally adequate and safe foods or limited or uncertain ability to acquire acceptable foods in socially acceptable ways."[5]

The above definition illustrates some key characteristics of FIS, firstly, it's a milieu or a social situation; secondly, having a finite amount of food ("uncertain availability"); thirdly, uncertain ability or inability to procure food ("unaffordable"); fourthly, lack of guaranteed access ("uncertain economic access") to sufficient quantity; and fifthly, lack of nutritionally adequate and safe food ("inadequate nutritionally and safety standard"). The literature demonstrates that many researchers attempting to explore the causes of FIS discussed these five (5) characteristics of FIS. Such as, reflecting the episodic nature of food insecurity, researchers asserted that food can accessed, but when the amount of food produced is inadequate FIS pertains.[6] Similarly, positive association between increasing food prices and increased prevalence

---

[2] FAO. Hunger and food insecurity [EB/OL] www.fao.org/hunger/en/ n.d/2022-03-14.

[3] FAO. Trade Reforms and Food Security: Conceptualizing the Linkages. [R] Rome: FAO, 2003. See also, Thomas, H. C. (eds.). Trade reforms and food security: country case studies and synthesis [M] Rome: FAO, 2006.

[4] FAO, IFAD. WFP, The State of Food Insecurity in the World 2013-The Multiple Dimensions of Food Security. [R] Rome: FAO, 2013.

[5] United States Department of Agriculture Economic Research Service. Definitions of food security. [EB/OL]. http://www.ers.usda.gov/topics/food-nutrition-assistance/food-security-in-the-us/definitions-of-food-security.aspx#ranges n.d/2022-03-14, See also, Christine M Olson, David H Holben. Position of the American Dietetic Association: Domestic Food and Nutrition Security [J] Journal of the American Dietetic Association, 2002, 102 (12): 1840–1847.

[6] Leblanc M, Kuhn B, Blaylock J. Poverty amidst plenty: Food insecurity in the United States [J] Agricultural Economics, 2005 (32):159–173.

of food insecurity is observed, which associated poverty with food insecurity.[7] Some researchers observed, association between sufficient quantity, inadequate nutritionally and safety standard of food with health and active living (Laraia et al. 2006; Zhang et al. 2013). Additionally, the milieu or a social situation of FIS is classified into three major categories by the level of severity:

1. **Acute:** Severe hunger and malnutrition to the point that lives are in a significant threat immediately (e.g. famine);
2. **Occasional:** It is defined as existence of food insecurity due to a specific temporary circumstance; and.
3. **Chronic:** the requirement of food needs is consistently or permanently under threat.

There categories of FIS depending on the level of severity are influenced by many external factors, such as socio-economic condition, food production, climate change, market instability, and so forth. Among the other factors, market stability is a controllable factor however, trade policy factors and political factors holds an impact on it. Such as, Tariff barriers at the national border are identified as the major cause of market instability because of food price that ultimately affects food security.[8] To understand how tariff affects price, affordability, and access to food, recent changes by Trump Administration in United States (US) can be a good example. In June 2018, the Trump administration introduced billions of dollars in new tariffs on Chinese imports; as a result, the price of Chinese goods got increased. On the other hand, in August, China announced a 25% tariff on $16 billion worth of U.S. goods including vehicles and crude oil in retaliation to the U.S. tariffs on $16 billion worth of Chinese goods.[9] Matt Gold, a former deputy assistant U.S. Trade Representative for North America under former President Barack Obama, commenting on the situation stated, "you're going to see higher prices passed on to consumers [...] almost immediately."[10] He added, "a U.S. importer pays the taxes to the customs duties or customs tariffs to the U.S. Treasury [...] Of course, that's going to effect the sale price [and] whatever price at which the exporter sells to the importer is going to lower, because the importer has to pay duties in addition to paying the purchase price."[11]

---

[7] See, Zhang Q, Jones S, Ruhm CJ, Andrews M. Higher food prices may threaten food security status among American low-income households with children [J] Journal of Nutrition, 2013(143):1659–1665. See also, Gregory CA, Coleman-Jensen A. Do high food prices increase food insecurity in the United States? [J] Applied Economic Perspectives and Policy, 2013 (35): 679–707.

[8] See, Bhagwati, J. Protectionism [M] In David R. Henderson (eds.) The Concise Encyclopedia of Economics. Indianapolis: Liberty Fund Inc., 2002.

[9] CNBC. China slaps 25% tariffs on $16 billion worth of US goods.[N/OL]. https://www.cnbc.com/2018/08/08/china-announces-25percent-tariffs-on-16-billion-worth-of-us-goods-including.html. 2018-08-08/2022-02-22.

[10] Gillies, T. Pricier soda and food: How tariffs are set to hit your wallet [N/OL]. https://www.cnbc.com/2018/07/27/how-trumps-tariffs-are-set-to-drive-food-and-drink-prices-higher.html 2018-07-27/2022-02-22.

[11] Washington, G. Why Americans Are About To Experience Sharply Higher Price [N/OL]. https://www.theburningplatform.com/2018/07/30/why-americans-are-about-to-experience-sharply-higher-prices/comment-page-1/ 2018-07-30/2022-02-22.

This indicates that if the price of foods is increased the consumers have to pay more. Subsequently, food might be available but not be affordable; so, food might not be accessed due to higher price. Hence, it is apprehended that, hunger and malnutrition is the global health problem which is caused because of market instability which is causing high price of food.

Food insecurity is a very complex phenomenon due to a range of factors that are involved, however, insufficient amount of literature exists which made an attempt to define and conceptualize FI. Most to the literature defines and conceptualizes food security (FS) and comprehends FI as antonym of FS. So, despite the multidimensional aspect and contrasting impact on every section of the population of FI, researchers and scholars have given more emphasis on FS, making FI antithesis of FS.

2.2.2 Food Security

FS is an evolving multidimensional concept. The term was first defined in 1974 World Food Conference in terms of food supply—assuring the availability and price stability as, "availability at all times of adequate world food supplies of basic foodstuffs to sustain a steady expansion of food consumption and to offset fluctuations in production and prices."[12] Reutlinger & Knapp defined FS as a condition in which the probability of a country's citizens falling below a minimal level of food consumption is low (Reutlinger 1985). FAO in the Director Generals Report (FAO 1982) focusing on food access defined FS as, "ensuring that all people at all times have both physical and economic access to the basic food that they need."[13] In mid-1980s, there was a shift in focus from availability, stability, and access to rights and entitlements influenced by Amartya Sen's seminal study *Poverty and Famines* (Sen 1983) which reflected in World Bank Report on Poverty and Hunger (World Bank 1986). However, Chambers and Conway livelihood approach toward FS which presented that a host of stresses can interact to affect food security at a household or individual level, (Chambers and Conway 1992) which have influenced the FAO World Food Summit's definition of FS in 1996 (WFS 1996) which states food security exists "when all people at all times have access to sufficient, safe, nutritious food to maintain a healthy and active life." Later in World Food Summit definition of FS addressed the multidimensional nature of food security. World Food Summit stated, "food security exists when all people, at all times, have physical and economic access to sufficient, safe and nutritious food that meets their dietary needs and food preferences for an active and healthy life."[14] So beside identifying FS as milieu or a social situation, like FIS, World Food Summit gave four pillars to assess FS, however, the

[12] FAO, An Introduction to the Basic Concepts of Food Security [R] Rome: FAO 2008

[13] FAO. Director Generals Report on World Food Security: A Reappraisal of the Concepts and Approaches [R] Rome: FAO, 1983.

[14] UN. Report of the World Summit on Sustainable Development, Johannesburg, South Africa. UN Doc. A/CONF.199/20 revised by UN Doc. A/CONF. 199/20/Corr. 1 [R] New York: UN, 2002.

standard of FS is not categorized depending on the level of FS. The four pillars to assess FS are: (WFS 1996)

1. **Food availability:** The availability of sufficient quantities of food of appropriate quality, supplied through domestic production or imports (including food aid).
2. **Food access:** Access by individuals to adequate resources (entitlements) for acquiring appropriate foods for a nutritious diet. Entitlements are defined as the set of all commodity bundles over which a person can establish command given the legal, political, economic and social arrangements of the community in which they live (including traditional rights such as access to common resources).
3. **Utilization:** Utilization of food through adequate diet, clean water, sanitation and health care to reach a state of nutritional well-being where all physiological needs are met. This brings out the importance of non-food inputs in food security.
4. **Stability:** To be food secure, a population, household or individual must have access to adequate food at all times. They should not risk losing access to food as a consequence of sudden shocks (e.g. an economic or climatic crisis) or cyclical events (e.g. seasonal food insecurity). The concept of stability can therefore refer to both the availability and access dimensions of food security.

Adhering to World Food Summit's definition of FS, FAO in 2006 defined the notion by referring it "to a condition where all people, at all times, have physical and economic access to sufficient, safe and nutritious food that meets their dietary preferences for an active and healthy life."[15] Additionally, the Committee on World Food Security (CFS) in 2009 illustrating the FS used the same definition. However, in 2009 FAO come up with a broader multidimensional approach toward FS and defined:

> Food security exists when all people, at all times, have physical, social and economic access to sufficient, safe and nutritious food that meets their dietary needs and food preferences for an active and healthy life. Household food security is the application of this concept to the family level, with individuals within households as the focus of concern.[16]

From the above definition of FS, it can be asserted that FS is not only about availability, accessibility, and affordability of food, it is also about consuming safe and nutritious food. It is noteworthy to mention, in the 2009 defining FIS, FAO stated "Food insecurity exists when people do not have adequate physical, social or economic access to food as defined above [in FS]." So present literatures indicate FS and FIS are multidimensional corresponding and inversing concepts, and one prevails over other.

---

[15] FAO, IFAD, UNICEF, WFP and WHO. The state of food security and nutrition in the world 2022 [R] Rome: FAO, 2022.

[16] CFS. Reform of the Committee on World Food Security Final Version. CFS:2009/2 Rev.2. [S] 2009: 1.

2.2.3 Food Safety

Food safety has become a shared concern among the countries as food trade expands, mostly, because of the impacts on human health. So, food safety is acknowledged as a major factor in ensuring good health. Hence, production of sufficient amount of safe and nutritious food is a global challenge faced by different stakeholders operating the global food system (Marvin et al. 2016). The literature suggests that food safety refers to a safety or security system establishing mechanism encompassing many facets of handling, preparation, and storage of food to prevent illness and injury (Marvin et al. 2016). So, food safety refers to the mechanism involved "in the preparation, handling and storage of food meant to prevent foodborne illness and injury,"[17] and holds no connection to access, availability, and affordability of physical, social, and economic access to sufficient quantity of food, rather shares direct connection with the safety aspects (i.e. safe and nutritious food) of access, available and affordable food that meets the dietary needs and food preferences required for an active and healthy life. Which mean, the food safety system is also a food security issue in terms of the consumption of safe and nutritious food. Hence, access to safe and nutritious food is considered a *basic individual right* (Uyttendaele et al. 2016) and *public health preventative measures* (Levy et al. 2022).

Food safety as an aspect of food security gained concern due to rapid growth of food trade (as discussed in Chap. 2) which stimulated "food crime" such as chemical contaminants, quality of food, misleading information on food label, and so forth. Although food safety is not only related to international food trade but, also to national food trade. According to WHO in the FAO Regional Conference for Asia and the Pacific (WHO 2018), "thirty-one foodborne hazards causing 32 diseases are included, being 11 diarrhoeal disease agents (1 virus, 7 bacteria, 3 protozoa), 7 invasive infectious disease agents (1 virus, 5 bacteria, 1 protozoan), 10 helminths and 3 chemicals. diarrhoeal diseases are the most common illnesses resulting from the consumption of contaminated food, causing 550 million people to fall ill and 230 000 deaths every year."[18] Furthermore, it is asserted that "diarrhoeal diseases are the most common illnesses resulting from the consumption of contaminated food, causing 550 million people to fall ill and 230 000 deaths every year."[19] This indicates the notion food safety relates to safety of human health, in other words, food safety relates to security of good and secure health (Havelaar et al. 2013). At present a number of public health preventative measures exists that governs the food safety system to food security in terms of food hygiene and food safety throughout the food production chain to prevent illness and injury, however, focuses mainly on food safety during international food trade. Such food safety system involves, establishing food

[17] SESOTEC. What is Food Safety? [EB/OL] www.sesotec.com/apac/en/resources/blog/what-is-food-safety 2020/2022-04-25

[18] WHO. WHO Initiative to Estimate the Global Burden of Foodborne Diseases WHO/FOS/15.02 [R] Geneva: WHO, 2015.

[19] WHO. Food Safety [EB/OL] www.who.int/news-room/fact-sheets/detail/food-safety 2020/2022-03-29.

safety standards promoted by the Codex Alimentarius (or "Food Code") the Hazard Analysis and Critical Control Points (HACCP), SPS Agreement, General Standard for the Labelling of Prepackaged Foods (1985), Guidelines on Nutrition Labelling (1985), Principles and Guidelines for the Exchange of Information in Food Safety Emergency Situations (1995), Accelerating efforts on food safety (2019), and so forth.[20]

The literature analysis suggests the notion "food safety" is not adequately defined by international organizations. Rather it is asserted that "food safety, nutrition and food security are inextricably linked. Unsafe food creates a vicious cycle of disease and malnutrition, particularly affecting infants, young children, elderly and the sick […] access to sufficient amounts of safe and nutritious food is key to sustaining life and promoting good health."[21]

2.3 Analyzing the Interconnectedness of Food Insecurity, Food Security, and Food Safety

From the above definitional analysis, it is comprehended that "FIS" is the antithesis of the notion "FS," however, "food safety" is an integral part of both the notion FIS and FS. Nevertheless, food safety is usually treated as a trade issue (Sykes 2006). To comprehend the nexus between FIS, FS, and food safety, FS is considered the focal point of the discussion (Fig. 2.1). The literature analysis of the contemporary international food law suggests that the foundation of the contemporary global food system is based on three major concepts, FIS, FS, and food safety. So, comprehending these three major concepts' the main aspects contributes to recognize the objective and nature of any food policy or international food law. Subsequently, it subscribes to developing and advancing international food law governing the global food system.

Starting the nexus analysis with the definition of FS, which states, "food security exists when all people, at all times, have physical and economic access to sufficient, safe and nutritious food that meets their dietary needs and food preferences for an active and healthy life."[22] Whereas, food insecurity exists "where some people do not have access to sufficient quantities of safe and nutritious food and hence do not consume the food that they need to grow normally and conduct an active and healthy life."[23] This indicates one situation prevails over the other, which means if FS prevails, FIS cannot exist, when measured in terms of sufficient quantity of food

[20] For more information regarding the food safety guidelines visit CODEX, Codex Texts: Guidelines [EB/OL]. http://www.fao.org/fao-who-codexalimentarius/codex-texts/guidelines/en/ n.d/2022-02-22.

[21] WHO. Food Safety [EB/OL] www.who.int/news-room/fact-sheets/detail/food-safety 2020/2022-03-29.

[22] FAO. An Introduction to the Basic Concepts of Food Security [EB/OL https://www.fao.org/3/al936e/al936e00.pdf n.d/2022-03-25.

[23] Simon, George-André. History of Food Security [EB/OL] http://www.fao.org/fileadmin/templates/ERP/uni/F4D.pdf. 2012/2020-10-16.

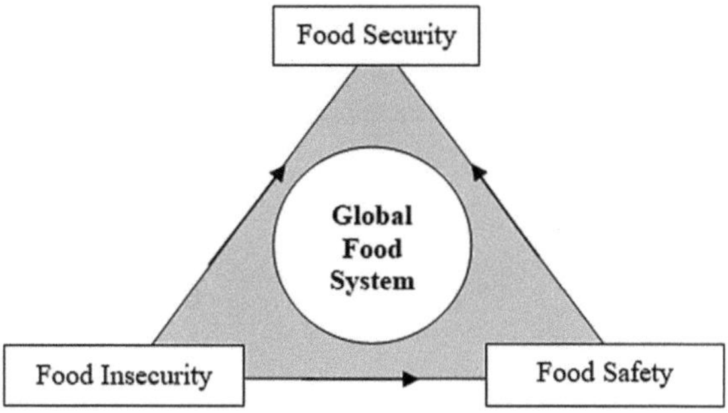

Fig. 2.1 Three concepts as the foundation of the Global Food System. (Source: Author Developed)

available, accessible, and affordable. However, contrasting situations might appear regarding food consumption, as both situations have emphasized on the consumption of safe and nutritious food. This means although a sufficient quantity of food is available, accessible, and affordable, food insecurity can exist if the food is not safe and nutritious food. So, the notion of food security holds two aspects, first, in term of availability, accessibility and affordability of sufficient quantity of food (i.e. food available, accessible, and affordable), and second, in term of consumption of safe food (i.e. food safety) and nutritious food (i.e. food standard) (Fig. 2.2); which is vice-versa in the case of FIS. This means, although food safety is a separate notion but cannot be represented or comprehended explicitly from the notion FIS and FS.

Furthermore, it is apprehended that, although the contemporary global food system circulates around the three notions FIS, FS, and food safety, it is observed that the equilibrium of the global food system can either shift toward FIS or FS, where food safety pretends to be the resultant force/equilibrium point (Fig. 2.3). Influenced by Marshall (1890), the declining side of the structure in Fig. 2.3 represents FS situation, the ascending side presents FIS situation and the resultant force/equilibrium point presents the food safety situation. The status of the food safety situation i.e. toward FS or FIS, can be determined by observing magnitude of the ascending side and the level of food safety situation can be observed by finding the difference between the magnitudes of ascending and declining side.

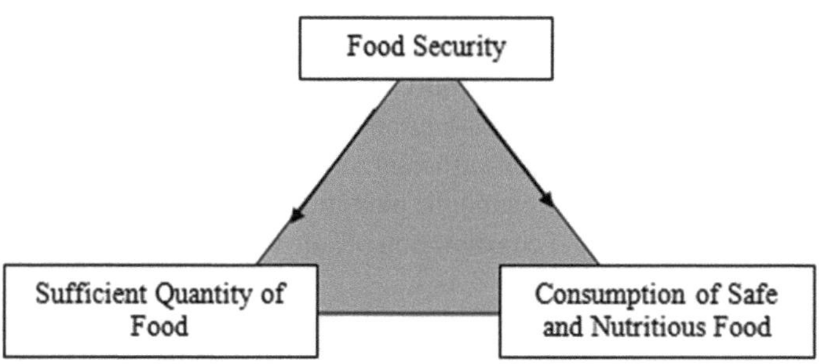

Fig. 2.2 Categories of food security. (Source: Author Developed)

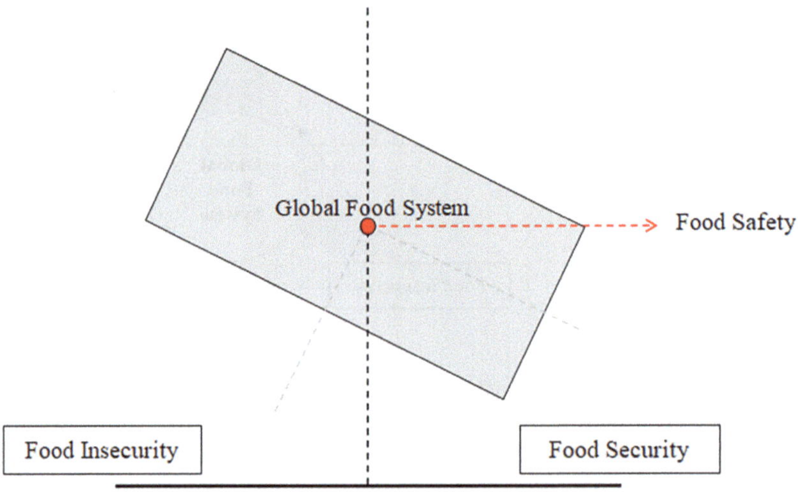

Fig. 2.3 Equilibrium of global food system. (Source: Author Developed)

The inextricable connection between FIS, FS, and food safety can further be understood by analyzing the measuring indicators to comprehend the standard of FS and FIS. Accordingly, the leading measuring indicators for FIS and FS situation depend on six (6) major indicators: availability, accessibility, affordability, utilization, nutritional standard, and safety standard of food (FAO n.d.) (Fig. 2.4) which can be measured to comprehend FIS and FS situation at different levels such as individual, household, national regional, and world. Hoddinott found approximately 450 food security indicators that exist (Hoddinott 2001). Such as dietary diversity score (DDS), food consumption score (FCS), Household Food Insecurity Access Scale (HFIAS), coping strategy index (CSI), calorie intake, and so forth (Carletto et al. 2013; Maxwell et al. 1999). Among all the food security indicators, definitional analysis of FIS and FS suggests six (6) major indicators which address the multidimensional aspect of the notion FIS and FS (Fig. 2.4), such as food supply chain, socio–economic, food hygiene, health, and so forth. Many researchers have attempted to develop a model to measure food security (Vhurumuku n.d.; Umaroh and Pangaribowo n.d.). However, the model does not address food security as a whole, i.e. in terms of availability, accessibility, and affordability of sufficient quantity of food meeting the dietary and nutrients need, and consumption of safe food. Most models focused on measuring food security in terms of sufficient food to meet dietary and nutrient needs. Notwithstanding, food security cannot be measured, without addressing availability, accessibility, affordability, and consumption of safe food, only from quantity and the dietary and nutrients perspective.

Additionally, it is observed that, although, there is a need to develop measure indicators for food safety, challenges are faced in developing measuring indicators for food safety, mostly because the notion is distinct from the notion FIS and FS, but yet connected (Fazil et al. 2008; Jaykus et al. n.d.; Luning et al. 2008; Serapiglia

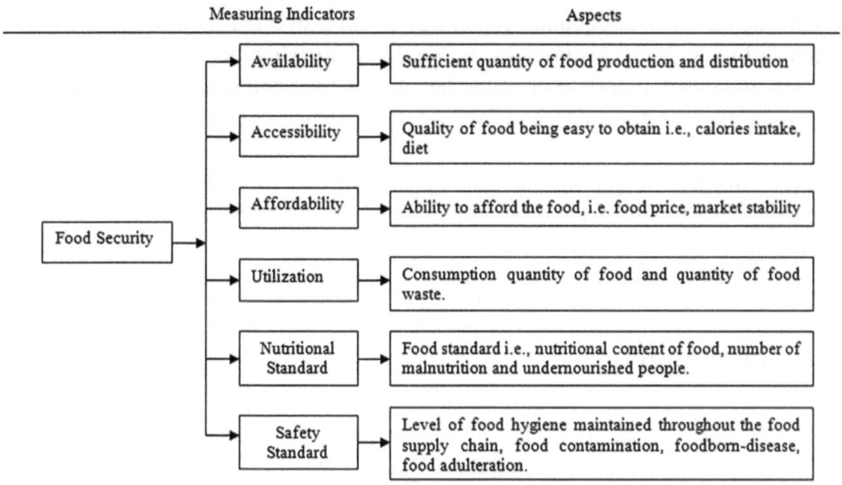

Fig. 2.4 Food security and food insecurity measuring indicator. (Source: Author Developed)

et al. 2007). Literature suggests that the study mainly focused on developing a food safety management system (Luning et al. 2008) instead of developing measuring indicators for food safety to comprehend the standard of food safety which indicates food security standards at the consumption level. It is asserted that FIS, FS, and food safety are inextricably linked, where food safety holds a smaller portion of FS when compared with the portion held by FIS. Although FIS and FS present contrasting milieus or a social situations or status of the global food system, however both are interconnected where one cannot be comprehended without comprehending the other notion and food safety holding a distinct position between the two notions acts as the center of attraction between FIS and FS, i.e. FIS $\leftrightarrow$ F $\leftrightarrow$ FS. Therefore, the policy and practice interventions need to consider the interrelationships between food insecurity, food security, and food safety and aim for synergies and co-benefits rather than trade-offs or conflicts. Some of the policy and practice implications are:

Adopting a holistic and integrated approach to address the multiple dimensions of food insecurity, food security, and food safety across the food system. This may involve strengthening coordination and collaboration among different actors and sectors involved in food production, processing, distribution, and consumption; harmonizing standards and regulations across different levels and regions; and enhancing monitoring and evaluation systems to track progress and impacts.

Applying a human rights-based approach to ensure that everyone has access to safe, sufficient, nutritious, and culturally acceptable food. This may involve recognizing the right to freedom from hunger as a fundamental human right; empowering vulnerable groups such as women, children, indigenous peoples, and smallholders to claim their rights; holding states and other duty-bearers accountable for fulfilling their obligations; and ensuring participation, transparency, accountability, non-discrimination, and rule of law in decision-making processes.

Leveraging innovation and technology to improve the efficiency, resilience, sustainability, and inclusiveness of the food system. This may involve investing in research and development to generate new knowledge and solutions for addressing food insecurity, food security, and food safety challenges; adopting digital technologies such as mobile applications, blockchain, artificial intelligence, and big data to enhance information sharing, traceability, transparency, and risk management; and supporting the scaling up and dissemination of innovative practices and technologies that can benefit smallholders, consumers, and the environment.

2.4 Conclusion

This chapter revisits the three key major notions, FIS, FS, and food safety, which forms the foundation of the global food system. Specifically, the definitional analysis of these three key major notions conceptualizes the notions provides a clear understanding about the notions and their inextricably linked. By providing the two aspects of the notion FS, the chapter asserted that FIS and FS presents contrasting milieu or a social situation or status of the global food system in terms of availability, accessibility, and affordability of food and food safety holding a distinct position presents FS and FIS milieu or a social situation in terms of consumption of safe food and safe health.

Furthermore, the definitional analysis of FIS, FS, and food safety contributes in identifying the targeted issues and effective ways for food policy interventions, which will support the analysis of the international food law in the subsequent chapters by identifying the objective of the law and which aspect of food security it focuses. This will allow policymakers to develop and enhance the contemporary international food law. Additionally, the chapter provided an equilibrium method to determine the status of global food system in terms of FIS, FS, and food safety, and a model to determine the standard of FS using measuring indicators which addresses the multidimensional aspect of FS. This contributes in identifying the major areas for policy intervention.

References

Anderson MD, Cook JT (1999) Community food security: practice in need of theory? Agric Hum Values 16(2). https://doi.org/10.1023/A:1007580809588

Carletto C, Zezza A, Banerjee R (2013) Towards better measurement of household food security: harmonizing indicators and the role of household surveys. Glob Food Sec 2(1):30–40. https://doi.org/10.1016/j.gfs.2012.11.006

Chambers R, Conway G (1992) Sustainable rural livelihoods: practical concepts for the 21st century, IDS discussion paper 296. Institute of Development Studies, Brighton

FAO (1982) Director-General's report on world food security: a reappraisal of the concepts and approaches. https://doi.org/10.3/JQUERY-UI.JS

FAO (2000) The elimination of food insecurity in the horn of Africa – summary. https://www.fao.org/3/x8530e/x8530e02.htm. Accessed 20 Oct 2020

FAO (n.d.) Measuring food safety – food safety technical toolkit for Asia and the Pacific

Fazil A, Rajic A, Sanchez J, Mcewen S (2008) Choices, choices: the application of multi-criteria decision analysis to a food safety decision-making problem. J Food Prot 71(11):2323–2333. https://doi.org/10.4315/0362-028x-71.11.2323

Havelaar AH, Cawthorne A, Angulo F, Bellinger D, Corrigan T, Cravioto A, Gibb H, Hald T, Ehiri J, Kirk M, Lake R, Praet N, Speybroeck N, de Silva N, Stein C, Torgerson P, Kuchenmüller T (2013) WHO initiative to estimate the global burden of foodborne diseases. Lancet 381. https://doi.org/10.1016/s0140-6736(13)61313-6

Hoddinott J (2001) Choosing outcome indicators of household food security. In: Food security in practice: methods for rural development projects

Jaykus L-a, Acuff GR, Busta FF, Dickson JS, Hollingsworth A, Marcy J, Ann Marie M (n.d.) Managing food safety: a systematic approach. Food Technology Magazine. https://doi.org/10.1111/1750-3841.16532

Laraia BA, Siega-Riz AM, Gundersen C, Dole N (2006) Psychosocial factors and socioeconomic indicators are associated with household food insecurity among pregnant women. J Nutr 136(1):177–182. https://doi.org/10.1093/jn/136.1.177

Levy N, Cravo Oliveira Hashiguchi T, Cecchini M (2022) Food safety policies and their effectiveness to prevent foodborne diseases in catering establishments: a systematic review and meta-analysis. Food Res Int 156. https://doi.org/10.1016/j.foodres.2022.111076

Luning PA, Bango L, Kussaga J, Rovira J, Marcelis WJ (2008) Comprehensive analysis and differentiated assessment of food safety control systems: a diagnostic instrument. Trends Food Sci Technol 19(10):522–534. https://doi.org/10.1016/j.tifs.2008.03.005

Marvin HJP, Bouzembrak Y, Janssen EM, van der Fels-Klerx HJ, van Asselt ED, Kleter GA (2016) A holistic approach to food safety risks: food fraud as an example. Food Res Int 89:463–470. https://doi.org/10.1016/j.foodres.2016.08.028

Maxwell D, Ahiadeke C, Levin C, Armar-Klemesu M, Zakariah S, Lamptey GM (1999) Alternative food-security indicators: revisiting the frequency and severity of coping strategies. Food Policy 24(4):411–429. https://doi.org/10.1016/s0306-9192(99)00051-2

Reutlinger S (1985) Food security and poverty in LDCs. Finance Dev 22(4)

Russell EJ (1953) The white man's dilemma: food and the future. Int Aff 29(4). https://doi.org/10.2307/2606038

Sen A (1983) Poverty and famines. Oxford University Press, Oxford. https://doi.org/10.1093/0198284632.001.0001

Serapiglia T, Kennedy E, Thompson S, De Burger R (2007) Association of food premises inspection and disclosure program with retail-acquired foodborne illness and operator noncompliance in Toronto. J Environ Health 70(1)

Sykes AO (2006) Domestic regulation, sovereignty and scientific evidence requirements: a pessimistic view. In: Trade and human health and safety. Cambridge University Press, pp 257–270. https://doi.org/10.1017/CBO9780511511325.011

Umaroh R, Pangaribowo EH (n.d.) IOP conference series: Earth and environmental science determinants of rural household food security in Indonesia: the case of protein-based food consumption. https://doi.org/10.1088/1755-1315/451/1/012038

Uyttendaele M, De Boeck E, Jacxsens L (2016) Challenges in food safety as part of food security: lessons learnt on food safety in a globalized world. Procedia Food Sci 6. https://doi.org/10.1016/j.profoo.2016.02.003

Vhurumuku E (n.d.) Food security indicators

WFS (1996) World Food Summit 1996, Rome declaration on World Food Security. World Food Security

WHO (2018) WHO estimates of the global burden of foodborne diseases: foodborne diseases burden epidemiology reference group 2007–2015.

World Bank (1986) The World Bank annual report. https://documents.worldbank.org/en/publication/documents-reports/documentdetail/992001468740700461/the-world-bank-annual-report-1986/2020-10-20. Accessed 20 Oct 2020

Zhang Q, Jones S, Ruhm CJ, Andrews M (2013) Higher food prices may threaten food security status among American low-income households with children. J Nutr 143(10):1659–1665. https://doi.org/10.3945/jn.112.170506

Chapter 3
The Nexus Between Food Security and Food Waste in the Contemporary Global Food System: A Critical Analysis

3.1 Introduction

The world population continues to grow, and it is expected to reach 9.7 billion in 2050 from 7.7 billion.[1] However, United Nation World Food Program (UN WFP) and FAO continues to claim that "enough food is produced today to feed everyone on the planet,"[2] although the number of hungry people worldwide has been gradually rising and there are more than 690 million hungry people around world (FAO 2020). According to the latest report SOFI (FAO 2021) nearly one-tenth of the world population—up to 820 million people went hungry in 2020. This raises question—if enough food is being produced to feed the global population, then why "food security" still remains as a "societal risk"? Critical interpretive synthesis (CIS) of contemporary literature and the facts indicates—firstly, there is lack of equity in distribution of food, in other words uneven geographic distribution of food, in food supply chain of the global food system; secondly, food might be available but not accessible and affordable by people because of economic instability; thirdly, *enough food is produced to feed everyone* does not apply to the present condition to feed the 7.7 billion people. According to reports, the earth is losing fertile soil at the rate of 24 billion tons a year[3] and there is growing water shortage and scarcity (Bongaarts 2021), this indicates toward a clandestine impact of climate change on the food production and food standard, specifically agri-food. Hence, there are still 840 million undernourished

[1] Growing at a slower pace, world population is expected to reach 9.7 billion in 2050 and could peak at nearly 11 billion around 2100 [N] UN DESA 2019-06-17.

[2] See, Can we feed the world and ensure no one goes hungry? [N] UN News 2019-10-03 See also, FAO. The State of Food and Agriculture 2019 [R] Rome: FAO, 2019. See also, FAO Enabling Agriculture to Contribute to Climate Change Mitigation (Submission to UNFCCC) [R] Rome: FAO, 2009; See also, FAO Food security and agricultural mitigation in developing countries: Options for capturing synergies [R] Rome: FAO, 2009.

[3] 24 billion tons of fertile land lost every year, warns UN chief on World Day to Combat Desertification [N] UN News. 2019-06-16.

S. S. Silvee, *The Law of Food Security: The Legal Framework for a Sustainable and Equitable Global Food System*, LITES - Legal Issues in Transdisciplinary Environmental Studies 5, https://doi.org/10.1007/978-981-95-3636-8_3

people in the world and according to the reports, the pandemic may have added between 83 and 132 million people to the total number of undernourished in the world in 2020 (Bongaarts 2021). And lastly, an enormous amount of food is wasted and lost in each and every stage of the global food supply chain. According to Food Loss Index (FLI) and the Food Waste Index (FWI) prepared by FAO, around 14% of food is waste during production[4]; the report does not include the retail and consumption level. *Food Waste Index Report 2021* (United Nation Environment Programme 2021), prepared by United Nations Environment Programme (UNEP), focusing on retail and consumption level asserts, "around 931 million tons of food waste was generated in 2019, of which 61 per cent came from households, 26 per cent from food service and 13 per cent from retail."[5] So, it can be deduced food waste is a major global issue in the food supply chain of the contemporary global food system which is clandestinely exacerbating food insecurity (Ishikawa 2014).

There is a complex interplay between food security and food waste, as both concepts are intricately linked to the sustainability and equity of our food system. While food security refers to the availability, accessibility, and affordability of nutritious food for all individuals, food waste occurs when edible food is discarded or lost during production, processing, distribution, or consumption. In this context, exploring the themes of food security and food waste is crucial to understand how we can ensure a sustainable and equitable food system that not only provides adequate nutrition to all but also minimizes waste and conserves resources. This chapter analyzes the research question: What is the relationship between food waste and food insecurity within the global food system? What are the main causes and consequences of food waste for food insecurity in the global food system? What are the policies and initiatives to reduce food waste and improve food security in the global food system? What are the implications of incorporating anti-food waste measures in building a sustainable future and achieving sustainable development? Addressing food waste is crucial for achieving several Sustainable Development Goals, particularly SDG 2: Zero Hunger and SDG 12: Responsible Consumption and Production. Reducing food waste not only contributes to alleviating hunger and malnutrition by increasing food availability and accessibility but also minimizes the environmental impacts of food production and distribution. By promoting sustainable consumption patterns and reducing food loss and waste along the entire food supply chain, we can enhance food security, protect natural resources, and create more sustainable and equitable food systems.

The chapter is divided into seven (7) segments including introduction and conclusion. The chapter begins with examining the relation between food security and food waste then the discussion focuses on the policies and initiatives. Later section explains the methodology in details and analyzing the findings of the meta-analysis of six (6) studies and quantifying food waste data and then analyzing the findings of web-based survey at restaurants which bespeak the gap in food waste data.

---

[4] FAO. The State of Food and Agriculture 2019 [R] Rome: FAO, 2019.

[5] UNEP. Food Waste Index Report [R] Nairobi: UNEP, 2021: p. 8.

3.2 The Relationship Between Food Security and Food Waste: Building a Sustainable and Equitable Food System

Food security and food waste are two sides of the same coin. On the one hand, food waste poses a significant threat to food security, as it can reduce the availability and affordability of food for vulnerable populations. Food insecurity and food waste are two critical issues that exist in the global food system, both of which have severe implications for food security. Food insecurity refers to a condition in which people do not have access to sufficient, safe, and nutritious food to meet their dietary needs. On the other hand, food waste refers to the loss or wastage of food that is still fit for human consumption. Moreover, food waste also has significant environmental and economic implications, as it contributes to greenhouse gas emissions, water depletion, and energy waste. Therefore, exploring the complex relationship between food security and food waste is crucial to identify strategies that can promote a sustainable and equitable food system that ensures food security for all while minimizing waste and maximizing resource efficiency.

To many, food waste might be a trivial issue that can exacerbate food insecurity. However, the growing concern about food waste is gradually portraying its significance as the considerable global challenges it possesses to food security[6] and climate change (Katajajuuri et al. 2014) are witnessed. Many studies such as Lipinski, et al. have underlined the significance of food waste and the need to reduce it for food security and sustainable global food systems (Lipinski et al. 2013). Xue, & Liu examining food losses and food waste (FLW) data for 84 countries and 52 individual years during 1933 and 2014 in 202 publications, found that per capita food waste generated from households rose with the increase of per capita gross domestic product (Xue and Liu 2019). The purpose of their study was "to clearly and comprehensively understand the existing global FLW data on their quality and availability," as it's "a prerequisite for tracking the progress toward the SDG Target 12.3 and the national FLW reduction goals, and evaluating the effect of relevant policies." So, the study has assessed the bibliometric characteristics of existing literature and methods of measurement. Similar to this study many researchers have researched focusing on food waste quantification and measurement, such as Parfitt et al. (2010), Hartikainen et al. (2019), Møller et al. (2014), Thyberg et al. (2015), Aschemann-Witzel et al. (2015), Porpino et al. (2015), Stancu et al. (2016), and Koivupuro et al. (2012). Besides providing valuable knowledge concerning the food waste data collection methods, these studies have not only presented the major gaps in the existing global FLW data and need to understand the true scale of food waste and its impacts, monitor and manage waste level to apprehend where, why, and how much food is wasted in the food chain. However, these literatures affirm the enormous quantity of food waste created in the global food supply chain from production to consumption. These literatures also

[6] The Economist Intelligence Unit (EIU). Food Loss and Its Intersection With Food Security [EB/OL]. https://www.eiu.com/public/topical_report.aspx?campaignid=foodloss14 n.d/2022-01-02.

confirm that in the harvest to post harvest and in the consumption stage of global food supply chain food is mostly wasted. And it can be deduced that irrespective of the causes if large amount of food is waste it exacerbates food insecurity.

The studies have also affirmed that there is no unified agreement on the definition of food waste. For instance, FAO defined food waste in the *Food losses and waste in the context of sustainable food systems* (FAO 2014) as "food appropriate for human consumption being discarded or left to spoil at consumer level – regardless of the cause and food losses as a decrease, at all stages of the food chain prior to the consumer level, in mass, of food that was originally intended for human consumption, regardless of the cause."[7] Some studies holding the similar perspective have continued to make a difference between food loss and food waste such as, Buzby et al. (2014) and Gustavsson et al. (2011). However, noteworthy to mention FAO has defined the diptych term *food loss and waste* (FLW) similar to the term food loss as "a decrease, at all stages of the food chain from harvest to consumption in mass, of food that was originally intended for human consumption, regardless of the cause."[8] So, it can be asserted that FAO is considering food waste also a kind of food loss. However, later FAO defined food waste in *The State of Food and Agriculture* (FAO 2019) as the "decrease in the quantity or quality of food resulting from decisions and actions by retailers, food service providers and consumers."[9] Whereas, food loss as "the decrease in the quantity or quality of food resulting from decisions and actions by food suppliers in the chain, excluding retailers, food service providers and consumers."[10] So, with this definition FAO has distinguished between food waste and food loss by asserting that food waste occurs at the retail and consumer levels and food loss occurs at along production and supply chains (including post-harvest).

Many researchers such as, Blichfeldt et al. (2015), Papargyropoulou et al. (2014), and Katajajuuri et al. (2014) have defined food waste in terms of edibility versus inedibility or avoidability versus unavoidability. However, the latest study conducted by UNEP in *Food Waste Index 2021* (United Nation Environment Programme 2021), defined "food waste is defined as food, i.e. any substance – whether processed, semi-processed or raw – drink, and any substance that has been used in the manufacture, preparation or treatment of food that is intended for human consumption, and the associated inedible parts removed from the human food supply chain."[11] It is comprehended that according to the present definition food waste includes both edible and inedible parts of food. After analyzing the definitions of the food waste, in this study the following definition of food waste is considered:

> Food waste refers to edible and inedible parts of food eliminated, lost or discarded from the food supply chain which as no longer required or intended for human consumption, regardless of the cause, decreases the quantity or quality of food in the global food system.

---

[7] FAO. Food losses and waste in the context of sustainable food systems [R] Rome: FAO, 2014.

[8] Ibid.

[9] FAO. The State of Food and Agriculture 2019. Moving forward on food loss and waste reduction.[R] Rome. FAO, 2019.

[10] Ibid.

[11] UNEP. Food Waste Index Report 2021 [R] Nairobi: UNEP, 2021: p. 9.

The relationship between food insecurity and food waste is complex and multifaceted. On the one hand, food waste contributes to food insecurity by reducing the amount of food available for consumption. On the other hand, food insecurity can lead to food waste as people may not have the means to store, preserve, or use food effectively. In low-income countries, food waste is more prevalent in the production and post-harvest stages, whereas in high-income countries, it is more common at the consumption stage. Food waste can exacerbate food insecurity by reducing the availability of food and increasing prices, particularly for low-income households.

Reducing food insecurity and food waste requires a comprehensive and coordinated approach that involves all stakeholders in the food system. Addressing food insecurity requires addressing the underlying causes of poverty, inequality, and conflict. This can be achieved through various measures, including social protection programs, investment in rural development, and improving access to education and healthcare. Reducing food waste requires interventions at every stage of the food supply chain, from production to consumption. This can include measures such as improving infrastructure, implementing better storage and transportation systems, and educating consumers on the importance of reducing food waste. The interconnectedness of food security, food insecurity, food safety, and food waste in the global food system highlights the need for a comprehensive and coordinated approach to addressing these issues. While significant progress has been made in addressing these issues, much remains to be done to ensure that all people have access to safe, nutritious, and sufficient food. Efforts to address these issues must be guided by an understanding of the underlying social, economic, and political factors that drive food insecurity and food safety issues, and must be informed by the latest research and data on these issues. Ultimately, addressing these issues will require a sustained and collaborative effort by governments, civil society organizations, the private sector, and individuals to ensure a more equitable and sustainable food system for all.

3.3 Policies and Initiatives to Reduce Food Waste and Improve Food Security in the Global Food System

In this section we analyze the relevant legal sources and principles for addressing food waste and food insecurity at the international, regional, national, and local levels. It identifies the main legal actors and stakeholders involved in the regulation and implementation of anti-food waste measures. It also evaluates the strengths and weaknesses of the current legal framework and its effectiveness and efficiency in reducing food waste and enhancing food security.

There are several legal sources and principles at international level that address food waste and food insecurity from human rights law, international development law, and international environmental law perspectives. For example, the UDHR and the ICESCR recognize the right to adequate food as part of the right to an adequate standard of living and the right to be free from hunger.[12] The right to food imposes obligations on states to respect, protect, and fulfill the availability, accessibility, adequacy, and sustainability of food for all people without discrimination. The Rome Declaration on World Food Security and the World Food Summit Plan of Action affirm the commitment of states to ensure food security for all people at all times and to reduce food losses and waste.[13] The SDGs, especially SDG 2 (Zero Hunger) and SDG 12 (Responsible Consumption and Production), set targets and indicators for reducing food waste and enhancing food security by 2030.[14] The Voluntary Guidelines to Support the Progressive Realization of the Right to Adequate Food in the Context of National Food Security provide guidance for states on how to implement the right to food and address food waste and food insecurity in their policies and programs.[15] The Voluntary Guidelines on Food Systems and Nutrition provide guidance for states on how to promote healthy and sustainable food systems that prevent and reduce food waste while ensuring food security and nutrition for all.[16] Although these international frameworks and initiatives are not legally binding, but, it provides guidance and support for countries to adopt and implement anti-food waste measures. For example, the United Nations Environment Assembly (UNEA) adopted a resolution on the prevention, reduction and reuse of food waste in 2019, which calls for actions at all levels to achieve the SDG target 12.3 to halve per capita global food waste at the retail and consumer levels by 2030 and reduce food losses along production and supply chains. The FAO launched the Save Food initiative in 2011, which aims to raise awareness and facilitate collaboration among different stakeholders to reduce food loss and waste. The FAO also developed the Food Loss Index (FLI) and the Food Waste Index (FWI) to measure and monitor the progress on food loss and waste reduction.

At the international level, the main legal actors involved in the regulation and implementation of anti-food waste measures include UN agencies such as the FAO, the WFP, the WHO, and the UNEP; intergovernmental organizations such as the WTO and the World Bank; regional organizations such as the African Union (AU) and the EU; non-governmental organizations (NGOs) such as Oxfam and Action Against Hunger; private sector actors such as multinational corporations and farmers' associations; and civil society actors such as consumers' groups and human rights defenders. The international legal framework has several strengths including its universal recognition of the right to food as a human right, its comprehensive coverage of various aspects of food waste and food insecurity, its promotion of cooperation and coordination among different actors and stakeholders, its provision of guidance

---

[12] Committee on Economic, Social and Cultural Rights. General Comment No. 12 The Right to Adequate Food E/C.12/1999/5. [S] 1999.

[13] World Food Summit Plan of Action [C] World Food Summit, Rome, 2019.

[14] United Nations. Sustainable Development Goals [S] 2015.

[15] Committee on World Food Security. Voluntary Guidelines to Support the Progressive Realization of the Right to Adequate Food in the Context of National Food Security [S] 2004.

[16] Committee on World Food Security. Voluntary Guidelines on Food Systems and Nutrition [S] 2018.

and support for states in implementing anti-food waste measures, and its monitoring and reporting mechanisms for assessing progress and challenges. However, it also has several weaknesses including its lack of binding force and enforcement mechanisms, its reliance on voluntary compliance and political will, its fragmentation and inconsistency among different legal sources and principles, its insufficient attention to specific contexts and needs of different countries, and its limited participation and accountability of non-state actors.

Similar to the international level at the regional level the legal sources and principles that address food waste and food insecurity are different perspectives. For instance, the EU adopted the Circular Economy Action Plan in 2020, which sets a legally binding target to reduce food waste by 50% by 2030 at the retail and consumer levels, as well as a monitoring framework to measure food waste levels. The EU also adopted a directive on waste in 2018, which requires member states to take measures to prevent food waste generation along the food supply chain, facilitate food donation and surplus food redistribution, ensure separate collection of biowaste, and promote awareness raising campaigns. In Africa, there is no regional legal framework on food waste prevention and reduction, but there are some initiatives that aim to address the issue. For example, the African Union Commission (AUC) launched the Continental Post-Harvest Loss Management Strategy in 2014, which provides a framework for action to reduce post-harvest losses in Africa. The AUC also established the African Post-Harvest Losses Information System (APHLIS) in 2009, which provides data and information on post-harvest losses of cereals in sub-Saharan Africa.

As the right to food is recognized as part of the right to life, health, dignity, and development in various regional human rights instruments. For instance, the African Charter on Human and Peoples' Rights and the Protocol to the African Charter on Human and Peoples' Rights on the Rights of Women in Africa[17]; the European Convention on Human Rights and the Charter of Fundamental Rights of the European Union in Europe; and the Inter-American Democratic Charter and the American Declaration on the Rights of Indigenous Peoples in the Americas. The Malabo Declaration on Accelerated Agricultural Growth and Transformation for Shared Prosperity and Improved Livelihoods commits African states to halve post-harvest losses by 2025 and enhance resilience of livelihoods and production systems to climate variability and other shocks.

The main legal actors at the regional level include regional organizations, such as the African Union (AU), the European Union (EU), the Economic Community of West African States (ECOWAS), the Southern African Development Community (SADC), etc.; regional courts and tribunals, such as the African Court on Human and Peoples' Rights, the European Court of Human Rights, the Court of Justice of the European Union, etc.; regional development banks, such as the African Development Bank, the European Investment Bank, etc.; regional NGOs, such as the African Network for the Right to Food (ANoRF), the European Food Banks Federation (FEBA), etc.; regional private sector actors, such as regional corporations, farmers'

[17] Protocol to the African charter on human and people's rights on the rights of women in Africa (Maputo Protocol) [S] 2003.

unions, etc.; and regional civil society actors, such as regional consumers' associations, human rights networks, etc. The strengths of the regional legal framework include its adaptation to the specific contexts and needs of different regions; its promotion of regional integration and cooperation on food waste and food insecurity issues; its provision of binding and enforceable legal instruments and mechanisms; its facilitation of cross-border trade and movement of food products; and its involvement of various regional actors and stakeholders. The weaknesses of the regional legal framework include its variation and divergence among different regions; its limited scope and coverage of some aspects of food waste and food insecurity; its potential conflict or inconsistency with international or national legal sources and principles; its lack of resources and capacity to implement and monitor anti-food waste measures; and its limited participation and accountability of some regional actors and stakeholders.

However at the national level, there are different legal approaches to food waste prevention and reduction among countries. Some countries have adopted specific laws or regulations on food waste prevention and reduction, such as France, Italy, South Korea, China, Japan, Taiwan, Peru, Costa Rica, Chile, Colombia, Mexico, Canada, Australia, and New Zealand, among others. These laws or regulations may include provisions on food donation incentives or obligations, food waste reduction targets or obligations, food waste reporting or monitoring requirements, food waste penalties or fees, food waste education or awareness raising programs, etc. Other countries have integrated food waste prevention and reduction measures into their broader legal frameworks on waste management or environmental protection, such as Germany, Sweden, Denmark, Norway, Finland, Netherlands, United Kingdom (UK), United States (US), Brazil, Argentina, India, and Thailand, among others. These measures may include provisions on waste hierarchy or prevention principles, waste management plans or strategies, waste collection or treatment systems, waste taxes or charges, waste prevention or minimization programs, etc. For instance, The Food Waste Reduction Act in Canada, which establishes a national target to reduce food waste by 50% by 2030 and requires the federal government to develop and implement a strategy to achieve this goal. The Law on Prevention of Food Waste in France, which prohibits supermarkets from throwing away or destroying unsold food and obliges them to donate it to charities or for animal feed. The National Food Security Act in India, which provides subsidized food grains to eligible households under the Targeted Public Distribution System and creates legal entitlements for food security. The Law on Donation of Surplus Food in Italy, which simplifies and incentivizes the donation of surplus food from food businesses to non-profit organizations for social purposes. The Zero Hunger Program in Brazil, which integrates various policies and programs to combat hunger and poverty, such as food assistance, school feeding, family farming, etc. In South Africa, the Department of Environment, Forestry and Fisheries and the Council for Scientific and Industrial Research have developed a guideline for food waste prevention and management, which provides information on the drivers and possible actions that can be taken to reduce food waste throughout the food supply chain. The guideline is voluntary and aims to raise awareness and encourage collaboration among stakeholders. In addition, the South African Food

Loss and Waste Voluntary Agreement was launched in 2021 as a platform for public–private partnership to commit to reducing food waste by 50% by 2030.

In the EU, the Action Plan for the Circular Economy sets a binding target to halve per capita food waste at the retail and consumer levels by 2030, and reduce food losses along the production and supply chains. The EU also provides a common methodology and indicators to measure food waste, as well as guidelines on food donation, date marking, and valorization of food waste. Moreover, the EU supports various initiatives and projects to prevent and reduce food waste, such as the EU Platform on Food Losses and Food Waste, the REFRESH project, and the FUSIONS project.

In China, the Anti-Food Waste Law was enacted in 2021 to regulate food waste in catering services, schools, government agencies, and other public institutions. The law imposes fines for excessive food orders, leftovers, or promotions that encourage overconsumption. The law also encourages food donation, recycling, and education on food waste prevention. Furthermore, the law supports the development of standards and technologies for food preservation, processing, packaging, transportation, and storage.

These legal approaches show some similarities and differences in terms of their scope, objectives, instruments, and implementation. For example, both South Africa and the EU adopt a voluntary approach that relies on awareness raising, information sharing, and stakeholder engagement. However, the EU has a more comprehensive and binding framework that covers all stages of the food supply chain and provides clear targets, indicators, and guidelines. China has a more focused and mandatory approach that targets mainly catering services and consumers and imposes strict penalties for food waste. These differences reflect the different levels of development, governance structures, cultural norms, and food waste patterns of each country or region. In Rwanda, the Land Husbandry, Water Harvesting, and Hillside Irrigation project has helped control erosion, intensify yields on existing land, and provide greater protection from droughts, thereby reducing food losses due to climate change. Under this program, maize yields increased 2.6 times between 2009 and 2018. In Senegal, the West Africa Agricultural Productivity Program has developed new high-yielding, early maturing, drought-resistant varieties of crops such as sorghum, millet, groundnuts, and cowpeas. These varieties are being widely disseminated to farmers and have raised yields by an average of 30%, even with less and more erratic rainfall. In Kenya, the Twiga Foods platform connects smallholder farmers to urban retailers through a mobile-based marketplace that offers competitive prices and timely delivery. This reduces food waste by eliminating intermediaries and ensuring consistent demand and supply. In Nigeria, the ColdHubs initiative provides solar-powered walk-in cold rooms for smallholder farmers, retailers, and wholesalers to store and preserve perishable foods. This extends the shelf life of fruits, vegetables, and other perishable foods from 2 days to 21 days.

These policies and initiatives demonstrate some of the ways that food waste can be reduced and food security can be improved in the global food system. However, more efforts are needed to scale up these solutions and address the underlying causes of food waste, such as poor infrastructure, inadequate storage facilities, low consumer awareness, and weak policy enforcement. Furthermore, more data and research are needed to measure the extent and impact of food waste and to identify the most effective interventions for different contexts.

The main legal actors at the national level include national governments, parliaments, courts, ministries, agencies, commissions, etc.; sub-national governments, such as states, provinces, municipalities, etc.; national NGOs, such as food banks, food recovery organizations, etc.; national private sector actors, such as food producers, processors, distributors, retailers, etc.; and national civil society actors, such as citizens, consumers, farmers, etc. The strengths of the national legal framework include its direct applicability and enforceability within a country; its responsiveness to the specific contexts and needs of different countries; its flexibility and adaptability to changing circumstances and challenges; its innovation and experimentation with different legal measures and solutions; and its mobilization and empowerment of various national actors and stakeholders. The weaknesses of the national legal framework include its variation and divergence among different countries; its limited scope and coverage of some aspects of food waste and food insecurity; its potential conflict or inconsistency with international or regional legal sources and principles; its lack of resources and capacity to implement and monitor anti-food waste measures; and its limited participation and accountability of some national actors and stakeholders.

3.4 Research Methodology

To achieve the aims of the study, this empirical research has used mixed methods research (MMR), a method supported by many researchers such as Maxwell (2022). The research method involves combating both qualitative and quantitative data collection and analysis approaches within the same study. Allison Shorten and Joanna Smith noted in *Mixed Methods Research: Expanding the Evidence Base,* "mixed methods research requires a purposeful mixing of methods in data collection, data analysis and interpretation of the evidence. [...] uses[ing] a single research paradigm, either quantitative or qualitative. Data are collected and analyzed using different methods within the same paradigm."[18] An MMR approach triangulates the finds, and provides contextual details on the effectiveness of the law. Hence both qualitative and quantitative data is used. Neuman noted in *Basics of Social Research: Qualitative and Quantitative Approaches*, "a study that combines both qualitative and quantitative data tends to be richer and more comprehensive."[19] The study follows exploratory sequential design, by collecting and analyzing qualitative

[18] Shorten, A., & Smith, J. Mixed methods research: expanding the evidence base [J] Evidence Based Nursing, 2017, 20(3):pp. 74–75. https://doi.org/10.1136/eb-2017-102699

[19] Neuman, W. L. Basics of Social Research: Qualitative and Quantitative Approaches, 7th ed. [M] Essex: Pearson, 2006: p. 167.

data and moving sequentially to quantitative data collection and analysis to explain and contextualize qualitative findings.

3.4.1 Qualitative Data Collection and Analysis

As Ritchie noted in his work, *Qualitative Research Practice A Guide for Social Science Students and Researchers,* "in applied research, the policy-making cycle is sometimes used to define the different types of research needed during the key stages of policy-making – that is formulation, implementation and appraisal."[20] Hence, applied qualitative method will help to evaluate international guidelines and law on food, national law and enactments on food, international reports and case reports, such as, Codex Code, Codex Principles, SPS Agreement, Niigata Declaration on APEC Food Security, US-Federal Food and Drugs Act,[21] US-Food Quality Protection Act,[22] Food Hygiene Law of the People's Republic of China (for trial implementation),[23] and so forth. Furthermore, concentration is also given on analyzing articles published in scientific journals as academic discourses, taken to represent historically and culturally based knowledge about food security and food law, along with its possible causes and consequences. Scrutinizing academic discourses on food security and food law and their relative strength presented the kinds of knowledge which dominate in society in a given period. Furthermore, it presents how food security is generally perceived, and how food law is governing the global food system.

3.4.2 Quantitative Data Collection and Analysis

For the purpose of the quantitative data are analyzed using two methods—meta-analysis and statistical analysis. The quantitative data collection focuses specifically on food waste. Hence, FAO's database on food loss and waste is used which includes more than 480 publications and reports from various sources (including governments, universities, and international governmental and non-governmental organizations) of 155 countries, and contains 31,651 data points for meta-analysis.[24] Additionally, a web-based survey was conducted to assess people's eating habit at restaurants and wasteful behavior. The survey was conducted and developed using the website

[20] Ritchie, J. and Lewis, J. Qualitative Research Practice A Guide for Social Science Students and Researchers [M] LONDON: SAGE Publications, 2003: p. 26.

[21] US, Federal Food and Drugs Act [S] 2014.

[22] US, Food Quality Protection Act [S] 1996.

[23] Food Hygiene Law of the People's Republic of China (for trial implementation) [S] 2021.

[24] The dataset underlying the meta-analysis is available at www.faoorg/food-loss-and-food-waste/flw-data.

WJX.CN. later disseminated through internet using different mediums. The detail of the data analysis in discussed below.

1. **Questionnaire Design**
   Exploratory close format questionnaires are developed targeting regular restaurants visitors to examine whether food waste at restaurants is a usual manner which can contribute in food insecurity. Social desirability approach is taken in designing the thirty-six (36) survey questionnaire which is divided mainly into four (4) segments—Demographic profile, Eating Manner at Restaurants, Awareness of Food waste as Global Issue, and Anti-food waste Law (Appendix B).
2. **Questionnaire Validation**
   For the purpose of the study, content and predictive validation technique is used for questionnaire validation to ensure all the specific items on the questionnaire address the corresponds the questionnaire desires to measure and the accuracy of the questionnaire by detecting whether respondents' answers on the questionnaire can predict certain behaviors of the respondents.
3. **Sampling Unit and Technique**
   The study used non-probabilistic sampling technique which includes the individuals who are most accessible such as students and professional. The voluntary response of the respondents, who regularly visits restaurants, are recoded not to test the hypothesis but to develop an initial understanding of a small or under-researched population.
4. **Sample Size**
   Keeping the confidence level 95%, with the margin of error of ±5% and an expected sample proportion of 50%, the sample size considered is three hundred eighty-three (383) respondents. Of which three hundred sixty-five (365) respondents participated.
5. **Sample Design**
   The sample design is developed considering Sir Ronald A. Fisher's principle of randomization, which eliminates systematic bias, obtains a representative sample from the population, and helps in distributing the unknown variation.
6. **Data Collection Source**
   The primary data is collected and accumulated by conducting a web-based quantitative survey using the website WJX.CN and the URL is disseminated using Wechat, LinkedIn and email to the respondents.
7. **Statistical Tools**
   To analysis the data statistical analysis method is used using IBM "Statistical Package for the Social Sciences" (SPSS) software (SPSS-22). To facilitate the analysis, the ideal statistical tool to utilize is descriptive frequency statistical analysis.

3.4.3 *Ethical Considerations*

As the methodology conducts a non-interventional study (e.g. surveys) which contains human participation, therefore, the study adheres to Declaration of Helsinki and ensures the application of good human research ethics, concretely, information and consent issues of the participants and the obligations under the Personal Information Protection Law of People's Republic of China.[25] Acknowledging the minimum possibility of coercion and pressure on participants; the survey is circulated with a detail but limited information of the study to ensure veracity and transparency, however, not deception. The study ensures individual confidentiality and anonymity rights are secured and that all the participants are treated fairly, equally, and respectfully. Moreover, the study considers all the necessary guidelines that make a research ethically valid and confirms that the findings derived from the research is trustworthy so they can be readily used as references for future empirical works. The study considers the obligation under the principle of non-maleficence, academic ethics, and its obligation to produce an original work. Hence, all the sources that will be used in the research will be properly acknowledged. This is also applicable to both direct and indirect quotes and information that are used in the research. And specific referencing styles that are academically specified will be followed for referencing the sources. The study ensures confidentiality and anonymity rights are secured.

3.5 Meta-Analysis and Statistical Analysis

The study has divided the analysis into two categories, where one contributes in detecting extend of food waste in the global food system and understand attitude of different countries and jurisdiction toward food waste at international level, other helps to understand individual attitude toward food waste at individualistic level. The detail of the data collection and analysis in discussed below:

3.5.1 *Meta-Analysis of Previous Studies on Food Waste*

To conduct this study FAO's database on food loss and waste is used which includes more than 480 publications and reports from various sources (including governments, universities, and international governmental and non-governmental organizations) of 155 countries, and contains 31,651 data points.[26] FAO's database is already in the public domain aims at sharing knowledge and at increasing the evidence-base, and the use of FAO's database does not require prior authorization and can be used by anyone. Considering the huge amount of data, the study focuses on 10 years, 2010 to 2020.

[25] Personal Information Protection Law of the People's Republic of China [S] 2021.

[26] The dataset underlying the meta-analysis is www.faoorg/food-loss-and-food-waste/flw-data

As FAO's data set holds data collected using eleven (11) categories of data collection method, this study considers only six (6) data collection methods—cases study, controlled experiment, modeled estimate, FAO's annual production questionnaire, survey, and literature. The study did not consider other categories of study to avoid the potential of overlapping of data. Furthermore, to avoid the potential of overlapping of data, the study used data aggregation technique supported by FAO's database on food loss and waste.

Furthermore, FAO has used box plot to graphically represent the result of meta-analysis in the study "the State of Food and Agriculture."[27] In this study the findings of meta-analysis of FAO's database are presented using line graphs as the purpose of the study is to detect extend of food waste across the global food supply chain. Although, the different data collection method, obstacles data comparison because of different measurement methodologies, but, the meta-analysis provides useful insights which helped this study to substantiate food waste exacerbates food insecurity by presenting how much and where—along the supply chain—various foods are wasted, and the imperativeness of formulating anti-food waste system through legal framework to build food security.

3.5.2 Statistical Analysis of Web-Based Survey on Food Waste in Food Service

The quantitative method is used in this study mainly focuses on confirming the finding of the qualitative analysis due to insufficient amount of research reports addressing food waste at restaurants by consumers—analyzing consumers' food waste behavior. The details of quantitative data collection, sample design, and so forth is discussed in 504 Sect. 3.4.2. To avoid the risk of a common method, bias data was gathered by targeting three hundred eighty-three (383) respondents, requested to participate in the survey, of which three hundred sixty-five (365) participated. Among which 72.06% were students who crossing the age of maturity (≈76.16%) and 27.94% were professionals. Of which, 52.60% were male and 47.40% were female According to educational standard most of the participants were postgraduate 74.25%. The demographic profile results are shown in Table 3.1.

Although the data sample collected is not sufficient to understand the approach of 7.7 billion people toward food waste, however, as the data analysis focuses on food waste in food service area at consumption level and also analyzes consumers attitude toward food waste and its impact, it helps to estimate and comprehend the extend of food waste in food service area at consumption level and consumers attitude toward food waste. The data analysis will induce further research in this area.

---

[27] See, FAO. Food Loss and Waste Database [R]. Rome: FAO, 2019. See also, FAO. Moving forward on food loss and waste reduction [R] Rome: FAO, 2019.

Table 3.1 Distribution of respondents

Measure	Items	Frequency	(%)
Gender	Male	192	52.69%
	Female	173	47.31%
Age	Between 18 ~ 25	153	41.94%
	Between 26 ~ 30	125	34.41%
	Between 31 ~ 40	78	21.51%
	Between 41 ~ 50	9	2.15%
Educational level	Undergraduate	94	25.81%
	Postgraduate	271	74.19%
Occupation	Student	263	72.04%
	Work	102	27.96%

Source Author Analysis

3.5.3 The Importance of Meta-Analysis and Statistical Analysis in Understanding Food Waste and Food Security in the Global Food System

In this study meta-analysis has been conducted to identify what amount food is wasted throughout the food supply chain, which is important for understanding the extent of the problem of food waste in the global food system. The results of this analysis provided insights into the areas that require targeted interventions to reduce food waste and improve food security. Moreover, estimating the percent of food waste generated by the restaurant visitors through a statistical analysis, such as an online survey of restaurant visitors, is also crucial. As it provides detailed information about the factors that contribute to food waste by restaurant visitors. For example, statistical analysis can identify the most commonly wasted foods, the reasons why visitors leave food on their plates, the demographic characteristics that are associated with higher levels of food waste, the attitude of people toward wasting food, and along with individual contribution people in wasting food. This information can help to inform policy and practice recommendations for reducing food waste in the restaurant industry.

By combining the results of the meta-analysis and the statistical analysis of food waste generated by the food service industry, a comprehensive understanding of the food waste problem in the global food system. This helps to identify the key areas that require targeted interventions to reduce food waste and improve food security.

3.6 Data Analysis and Discussion

3.6.1 Food Waste in Global Food Supply: Results of Meta-Analysis

For the purpose of the study, the meta-analysis of different studies focusing on food waste across the contemporary global food system has between conducted separately depending on the six (6) data collection methods: cases study, controlled experiment, modeled estimate, FAO's annual production questionnaire, survey, and literature. These studies focused on food waste of different agri-food commodities, such as rice, mango, tomatoes, wheat, and so forth, at a particular stage of the food supply chain, such as, pre-harvest, post-harvest, harvest, farm, transport, processing, retail, households, storage, wholesale, and trade. However, some studies focused on the whole food supply chain. For the purpose of the study, total average percent of different agri-food waste at different stages of the food supply chain identified in these studies are calculated and considered as average food waste in the whole food supply chain for a particular year. The results of which is later combined to detect extend of food waste across the contemporary global food system and measures it under the "food waste severity measuring scale"[28] below (Fig. 3.1).

For measuring what percentage of food waste can exacerbate food insecurity and to control the level of food waste, the study developed food waste severity measuring scale which can be used by countries and international organizations to measure and control the level of food waste both nationally and internationally. Food waste severity measuring scale is a percentage measure of wasted food throughout the food supply chain which describes how impactful is the level of food waste on food insecurity, where a *iFw* below 15 is neutral, having no impact on food insecurity, and a *iFw* greater than 15 is holds moderate impact on food insecurity. The values of *iF* range from 5 to 50, where a value greater than 15 presents critical to severe impact on the food insecurity. The *iFw* is an abbreviation for "impact of food waste," where "*i*" is short for the word impact and "*Fw*" is the short for food waste. So *iFw* value will the total or average amount of food wasted in the food supply chain.

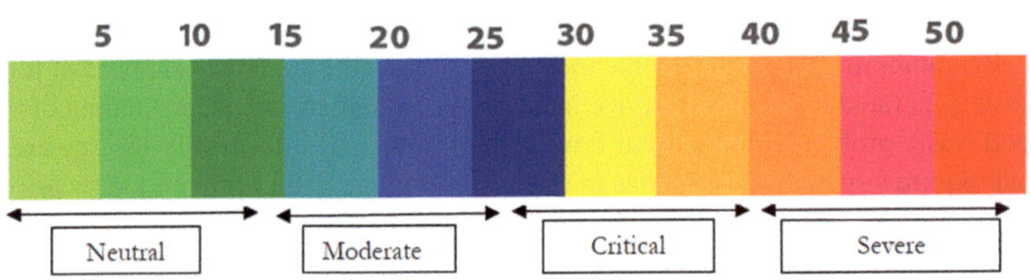

Fig. 3.1 Food waste severity measuring scale. (*Source* Author Developed)

[28] "Food waste severity measuring scale" is solely developed by the author by studying the pH scale to measure the percentage of global food waste which can exacerbate food insecurity.

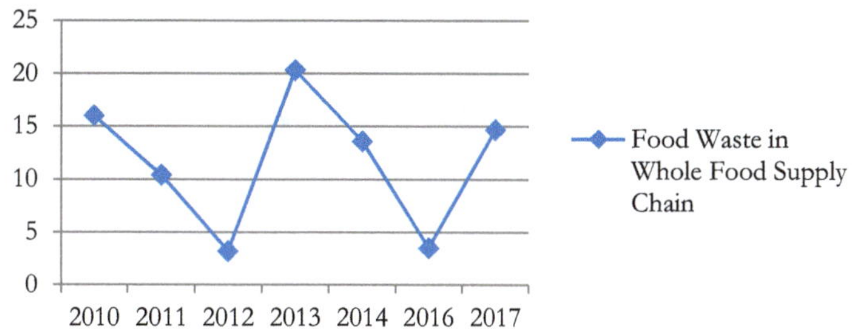

Fig. 3.2 Case study of food waste (between 2010 and 2017). (*Source* Author Analysis)

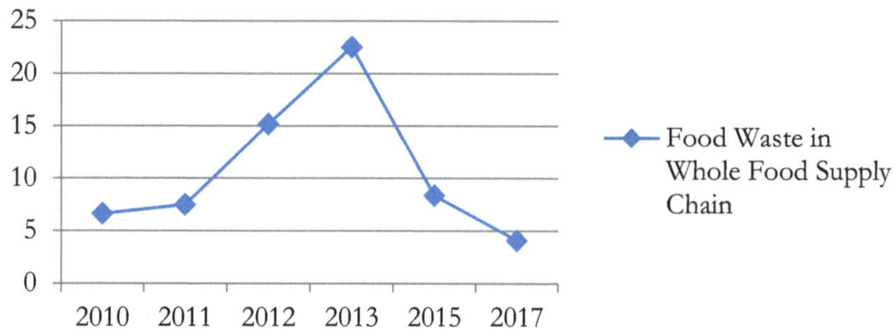

Fig. 3.3 Controlled experimental studies of food waste (between 2010 and 2017). (*Source* Author Analysis)

Starting with cases studies, the meta-analysis of studies from 2010 till 2020 (Fig. 3.2), finds that case study data of nineteen (19) countries[29] are available. The value Fig. 3.9 provides that the food waste value across the global food supply chain fluctuates between the ranges of 36% to 51%, which indicates toward critical to severe food waste condition.

The graph does not present the average percent of food waste of the year 2015 because of insufficient amount of case studies. However, Fig. 3.1 provides an overview of the meta-analysis of case studies on food waste and estimates the average percent of food waste, from pre-harvest to trade, in the whole food supply chain, which is approximately 13.04 percent per year. Further, showing the trend of food waste, Fig. 3.2 reveals there is a continual decrease in food waste from the year 2010 to 2012 and a peak increase in 2013 followed by a continual decrease and a steady increase in 2017. However, the meta-analysis of controlled experiment studies conducted in eight (8) countries[30] suggests a contrasting scenario in Fig. 3.3.

[29] The countries are Sub-Saharan Africa, Benin, Finland, Ghana, Haiti, India, Australia, Indonesia, Kenya, Republic of Korea, Malawi, Mexico, Nepal, Nigeria, Norway, Rwanda, Senegal, Sierra Leone, and Uganda.

[30] The countries are Finland, Ghana, Pakistan, United Republic of Tanzania, Sweden, Norway, Malawi and Kenya.

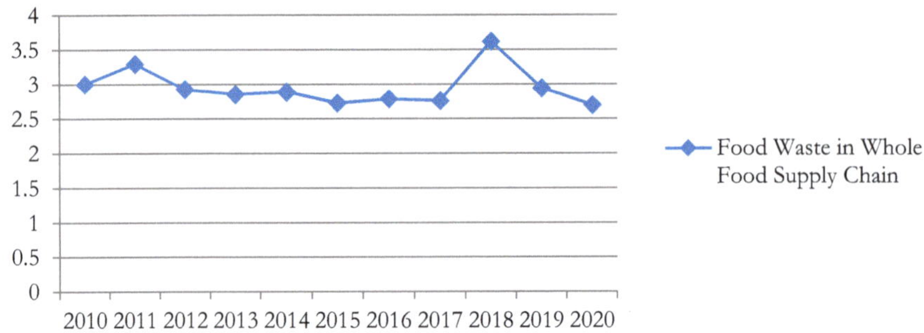

Fig. 3.4 Modeled estimate studies on food waste (between 2010 and 2020). (*Source* Author Analysis)

Figure 3.3 does not present the average percent of food waste of 2 years 2014 and 2016 because of insufficient data of controlled experiment studies. Figure 3.3 provides the meta-analysis of controlled experiment studies which presents a wide range of value between 4.06% and 22.50% and estimates the average percent of food waste, which is approximately 10.70 percent per year. Figure 3.3 provides that controlled experiment studies suggest a contrasting scenario when compared with case studies finding. It suggests there is an increase in food waste in the year 2012 and 2017. And there is a slight decrease in average food waste percent per year when compared with the average percent of food waste per year estimated by case studies. Furthermore, it suggests a trend of increase in food waste between the year 2010 to 2013. Similarly, the meta-analysis of modeled estimate studies suggests an incongruity range of value between 2.693% and 3.618% and estimates 2.953% as the average percent of food waste per year (Fig. 3.4).

Moreover, Fig. 3.4 provides that there was a fluctuation in food waste between 2010 and 2017 and then there was a slight increase in 2018 with continual decrease between the 2019 to 2020. Similarly, the meta-analysis of FAO's annual production questionnaire studies suggests a range of values contrasting not only to the value range presented by modeled estimate studies, but also to the other studies considered in this study (Fig. 3.5).

Figure 3.5 suggests an incongruity range of food waste value between 3.570% and 5.523% and estimates 4.727% as the average percent of food waste per year. Figure 3.5 provides a different trend in food waste over the years. It shows there is a continual decrease between the years 2010 to 2013 and increase between 2013 and 2016 and repeat of the trend by slight decrease from 2016 to a slight increase in 2018.

Figure 3.6 which presents the meta-analysis of survey studies on food waste suggests that the range of food waste value between is 2.489% and 15.369% and estimates 9.49% as the average percent of food waste per year. Figure 3.6 shows a steady trend in food waste which got a sharp decrease in 2019. Similarly, the meta-analysis on literature shows a different trend in food waste (Fig. 3.7).

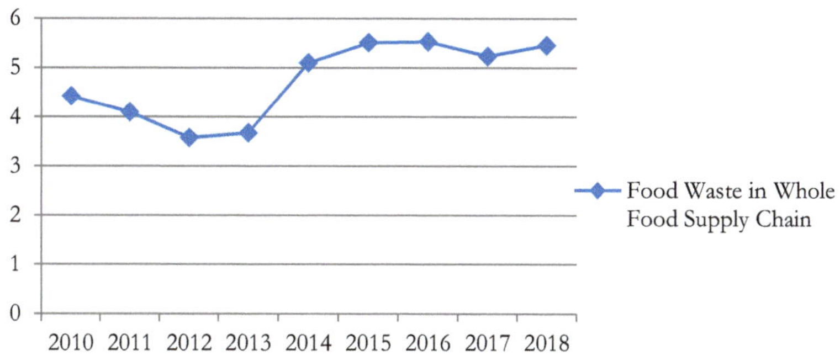

Fig. 3.5 FAO's annual production questionnaire studies on food waste (between 2010 and 2018). (*Source* Author Analysis)

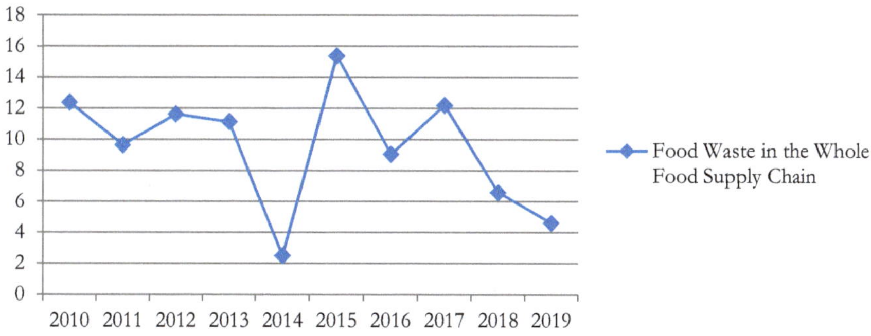

Fig. 3.6 Survey studies on food waste (between 2010 and 2019). (*Source* Author Analysis)

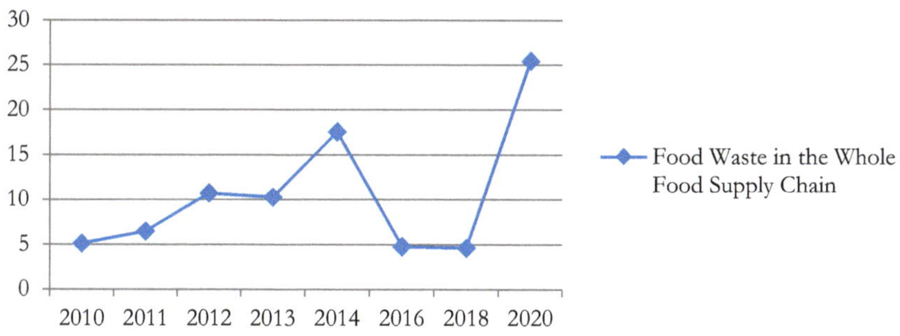

Fig. 3.7 1Literature on food waste (between 2010 and 2020). (*Source* Author Analysis)

Meta-analysis of literature on food waste suggests the range of food waste value between 4.57% and 25.4% and estimates 10.60% as the average percent of food waste per year. Figure 3.7 does not present the average percent of food waste of 2 years 2014 and 2016 because of insufficient literature studies, however, provides

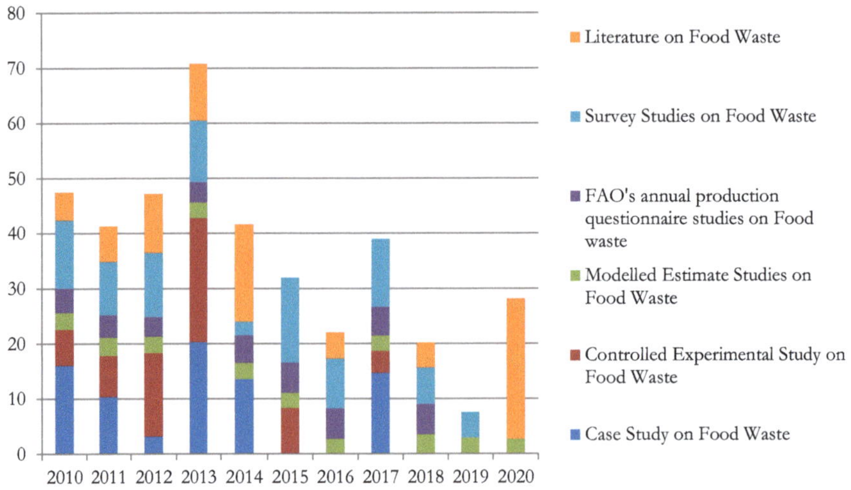

Fig. 3.8 Average estimated percent of food waste (between 2010 and 2020). (*Source* Author Analysis)

there was a continual fluctuation in food waste between 2010 and 2014 and increase in 2020.

For the purpose of the study, which is to detect the extend of food waste across the global food supply chain, the meta-analysis findings of six studies are combined in Fig. 3.8 which suggests that the average food waste across the global food supply chain per year is around 36%. However, sum-up of the average estimated percent of food waste in per year from the six studies in Fig. 3.9 suggests that the food waste across the global food supply chain in a year is around 51%.

Measuring both the values under food waste severity measuring scale suggests, the values indicate toward a food waste situation which is critical and sever with the *iFw* value of 36 and 51 (see Fig. 3.1). So, if 36–51% of food is wasted across the global food supply chain then approximately only 49–64% of food is consumed. This result clearly indicates that the relationship between food waste and food insecurity within the global food system is a complex and multifaceted issue. The meta-analysis of previous studies on food loss and waste based on FAO's database finds that there is insufficient amount of data measuring food waste in food services.[31] However, it provides that the food waste across the various stages of global food supply it is observed that enormous amount of food is wasted, firstly, at the farm level, i.e. agricultural level, pre-harvest, post-harvest level and storage; secondly, at processing and distribution, i.e. trade, transport and so forth; and lastly, at consumer level, i.e. household, food service and so forth. The findings of the meta-analysis substantiate the need of a comprehensive, harmonized, and standardized anti-food waste system at national level making an impact at international level.

[31] The dataset underlying the meta-analysis is available at www.fao.org/food-loss-and-food-waste/flw-data 2022-01-11.

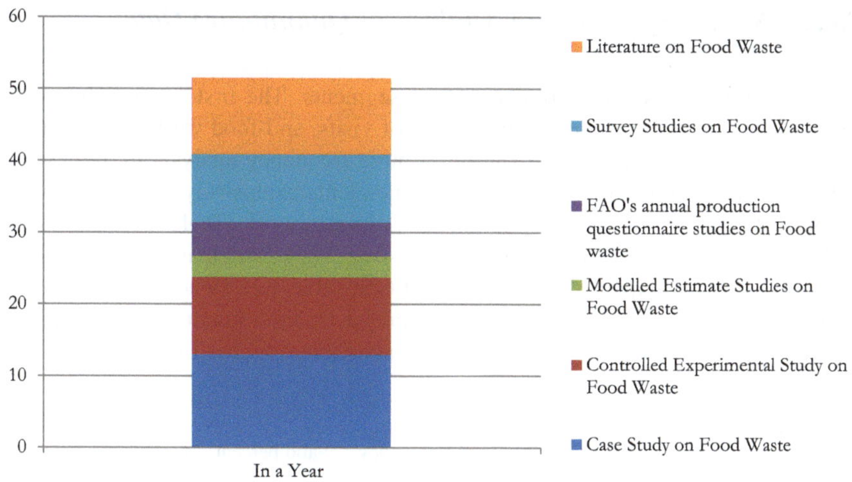

Fig. 3.9 Estimated total food waste in a year. (*Source* Author Analysis)

So, it can be stated that food waste and food insecurity are interrelated in a number of ways. For instance, food waste can lead to food insecurity by reducing the availability of food for those in need. Additionally, food waste contributes to food insecurity by reducing the efficiency of the food system, as resources are wasted in the production and distribution of food that is ultimately thrown away. Furthermore, food waste can exacerbate poverty, as resources that could have been used to meet the basic needs of the population are being wasted. The result highlights the relationship between food waste and food insecurity within the global food system is critical and provides important indications toward policymaking and the development of a legal and institutional foundation for anti-food waste measures to reduce global food waste and ensure a sustainable utilization of food. Addressing this relationship will require a multi-stakeholder effort, including governments, NGOs, businesses, and individuals, working together to create a more sustainable and equitable food system.

However, at consumer level, maximum studies focused on consumers' behavior at household level. Hence, to understand whether the consumers' behavior at food service level, in particular at restaurants, in the below section of the chapter focus is given on food waste behavior of consumers at restaurants—mostly due to less quantity of resources, the study has conduct specific survey focusing on food waste by consumers at restaurants.

3.6.2 Results of Statistical Analysis of Quantitative Data

The interpretation is categorized into four segments. The first section shows the respondent's visit at restaurants, frequency of visits, and food waste at restaurants (Tables 3.2, 3.3, 3.4, and 3.5). The descriptive frequency analysis in Tables 3.2, 3.3, and 3.4 suggests that 94.6% of the respondents extensively visits restaurants (Table 3.2) and 84.9% visits restaurants 1–7 times a week (Table 3.3). Tables 3.2 and 3.3 suggests that of three hundred sixty-five (365) participants, three hundred fifty-five (355) participants' visits restaurant (see Table 3.2). Hence, to maintain the accuracy of the analysis the sample size is decreased to three hundred fifty-five (355). Of which 70.6% of respondents wastes food at different times (Table 3.4).

Table 3.2 Frequency of restaurant visitors

	Frequency	Valid percent	Cumulative percent
Often	134	36.7	36.7
Generally	212	58.1	94.8
Never	19	5.20	100.0
Total	365	100.0	

Source Author Analysis

Table 3.3 Frequency of visit at restaurants

	Frequency	Valid percent	Cumulative percent
Once a week	204	55.9	55.9
2–4 times a week	94	25.9	81.8
5–7 times a week	12	3.3	85.1
More than 7 times	35	9.5	94.6
Never	20	5.4	100.0
Total	365	100.0	

Source Author Analysis

Table 3.4 Frequency of food waste at restaurants

	Frequency	Valid percent	Cumulative percent
Never	105	29.5	29.5
Always	18	5.1	17.6
Often	44	12.5	70.6
Occasionally	188	53.0	100.0
Total	355	100.0	

Source Author Analysis

Table 3.5 Witnessed food waste by others in the restaurant

	Frequency	Valid percent	Cumulative percent
Never	16	4.5	4.5
Always	113	36.4	69.4
Often	117	33.0	95.5
Occasionally	93	26.1	100.0
Total	355	100.0	

Source Author Analysis

The results of this descriptive frequency statistics are used to estimate the percent of respondent wasting food at restaurants using the below formula[32]:

$$\hat{y} = \left(\frac{P_w}{P_r}\right) * 100 \tag{3.1}$$

where:

- $\hat{y}$ is the estimated percentage of people wasting food at restaurants.
- P_w is the number of respondents who reported wasting food at restaurants.
- P_r is the total number of respondents who reported visiting restaurants.

Plugging in the values, we get:

$$\begin{aligned}\hat{y} &= \tfrac{70.6}{94.8} \times 100\\ &= 74.4\end{aligned}$$

So, the results of this descriptive frequency analysis estimate the rate of respondent wasting food at restaurants is $\geq$74.4%.

In Table 3.5, analyzing whether the respondents have witnessed others to waste food at restaurants, it is observed that 95.5% have witnessed food waste by others. The descriptive frequency analysis denotes a continuous and regular phenomenon of wasting food at restaurants.

The second section of the analysis focusing on to estimate how much food is wasted at restaurants at consumption level. Table 3.6 shows 59.1% respondents accepted waste of food ($\approx$) 100 g in each visit. According to sources, a person can intake ($\leq$) 2472 g (5.46 pounds) of food each day,[33] the quantity of food wasted by 210 respondents can be estimated to be sufficient to feed at least twenty-six (26) people once a day. In Table 3.6, it is observed 40.9% of respondents responded as unaware of the quantity of their food wasted, however, comparing the data with

[32] This equation relies on self-reported data from a survey, and may not reflect the actual percentage of restaurant customers who waste food.

[33] Writer, Staff. How Many Pounds of Food Does the Average Adult Eat in a Day? [EB/OL] www.reference.com/world-view/many-pounds-food-average-adult-eat-day-3f49d34cd3d872cd 2020-03-26/2020-04-18.

Table 3.6 Extend of food waste at restaurant

	Frequency	Percent	Cumulative percent
Unaware/nothing	145	40.9	40.9
≤100 g	198	55.7	59.1
≥100 g	12	3.4	100.0
Total	355	100.0	

Source Author Analysis

frequency of food waste at restaurants (Table 3.4) suggests 29.5% of respondents don't waste food at restaurants. Analyzing the response of these 29.5% respondents who claimed not to waste food at restaurants in Table 3.4, it is observed that 23.0% (≈31) of respondents agreed to waste ≤100 g of food at restaurants in each visit in Table 3.6. Of which 33.3% (≈10) respondents never take-away leftover food, 16.7% (≈5) respondents always take-away leftover food, and 50% (≈16) respondents once in a while take-away leftover food. So, Table 3.7 indicates 83.3% (≈26) respondents' waste ≤100 g of food at restaurants. This increases the frequency of food waste at restaurants by respondents presented in Table 3.4 from 70.6% (≈250) respondents to 77.7% (≈276) respondents. Here it is important to mention the few studies such as Brook Lyndhurst,[34] Neff et al.,[35] Hebrok and Boks,[36] Roodhuyzen et al.,[37] and Schanes et al.[38] According to which people are usually unaware of the amount of food they discard or waste because they lack the capacity to track, and understand the impact of their wasteful behavior, and many believe they waste less than other people do.

So, the comparative analysis of Tables 3.4, 3.6, and 3.7 suggests 77.7% (≈276) of 355 respondent wastes (≈) 100 g food at restaurants in each visit, the quantity of food wasted is supposed be enough to feed approximately twenty-six (26) people once a day. This increasing the estimated rate of respondent wasting food at restaurants to ≥81.9% from ≥74.4%. Considering the deficiency and other factors which might had an impact on the estimated value, the study considers estimated rate of respondents wasting food at restaurants as ≥74.4%.

34 Brook Lyndhurst. Food Behaviour Consumer Research: Quantitative Phase [EB/OL] https://www.wrap.org.uk/sites/files/wrap/Food%20behaviour%20consumer%20research%20quantitative%20jun%202007.pdf 2007-06-20/2022/01/21.

35 Neff, R.A., M.L. Spiker, and P.L. Truant. Wasted food: U.S. consumers' reported awareness, attitudes, and behaviors [J] PLoS ONE, 2015, 10(6).

36 Hebrok, M., and C. Boks. Household food waste: Drivers and potential intervention points for design: An extensive review [J] Journal of Cleaner Production, 2017 (151):380–392.

37 Roodhuyzen, D.M.A., P.A. Luning, V. Fogliano, and L.P.A. Steenbekkers. Putting together the puzzle of consumer food waste: Towards an integral perspective [J] Trends in Food Science & Technology 2017(68):37–50.

38 Schanes, K., K. Dobernig, and B. Gözet. Food waste matters—A systematic review of household food waste practices and their policy implications [J] Journal of Cleaner Production, 2018(182):978–991.

Table 3.7 Food waste habit of thirty-one (31) respondents agreed to waste ≤100 g of food attitude toward take-away of leftover food

	Frequency	Percent	Cumulative percent
Never	10	33.3	33.3
Once in a while	16	50.0	83.3
Always	5	16.7	100.0
Total	31	100.0	

Source: Author Analysis

In order to identify the percent of people wasting food at restaurants the below equation is considered[39]:

$$\hat{y} = \left(\frac{P_w}{P_r}\right) * \left[(p_1 * w) + (p_2 \times w)\right] * \left(\frac{po_w}{100}\right) \tag{3.2}$$

where:

- $\hat{y}$ is the estimated percentage of people wasting food at restaurants
- P_w is the number of respondents who reported wasting food at restaurants
- P_r is the total number of respondents who reported visiting restaurants
- p_1 is the percentage of respondents who reported visiting restaurants once a week
- w is the respondents who waste food $50 \leq$ value ≤ 100 g
- p_2 is the percentage of respondents who reported visiting restaurants 2–4 times a week
- po_w is the percentage of respondents who reported witnessing food waste by others.

Plugging in the values, we get:

$$\hat{y} = \left(\tfrac{70.6}{94.8}\right) * \left[(55.9{*}0.591) + (25.9{*}0.591)\right] \times \left(\tfrac{95.5}{100}\right)$$
$$= 22.51$$

The solution to the equation is 22.51, which represents the estimated percentage of people wasting food at restaurants. So, it is found that 22.51% of people visiting restaurant wastes food in each visit.

Although the findings of Table 3.8 suggest 77.3% of the 355 respondents addressed food waste as a serious problem. Table 3.9 shows 97.8% of the total 365 respondents agreed food waste exhausts resources (49.5%), exacerbates food insecurity (21.5%), and have environmental consequences (19.3%). Providing an in-depth to the respondents understanding about the impact of food waste, in the third section, Tables 3.10, 3.11 and 3.12 indicates that 59.1% agreed food waste can exacerbates national food

[39] This equation relies on self-reported data from a survey, and may not reflect the actual percentage of restaurant customers who waste food. Additionally, the equation assumes that the respondents who reported wasting food and visiting restaurants are representative of the overall population of restaurant customers, which may not be the case.

security (Table 3.11), 75.3% agreed food waste affects the sustainable development of a country and the world adversely (Table 3.10), and 72.0% agreed food waste holds adverse effect on social atmosphere (Table 3.12).

The results of the study denoted that out of 94.6% of people visiting restaurants, depict acknowledging that food waste exacerbates food security, 77.7% (≈276

Table 3.8 Severity of food waste

	Frequency	Valid percent	Cumulative percent
Very serious problem	274	77.3	77.3
Not a serious problem	28	8.0	22.7
Food is common	53	14.7	100.0
Total	355	100.0	

Source Author Analysis

Table 3.9 Impact of food waste

	Frequency	Valid percent	Cumulative percent
Waste of recourses	181	49.5	49.5
Environmental pollution	71	19.3	68.8
Exacerbates FS	78	21.5	90.3
Others	27	7.5	97.8
No effects	8	2.2	100
Total	365	100.0	

Source Author Analysis

Table 3.10 Impact of food waste on sustainable development

	Frequency	Valid percent	Cumulative percent
Disagree	47	12.9	12.9
Agree	275	75.3	87.1
Neutral	43	11.8	100.0
Total	365	100.0	

Source Author Analysis

Table 3.11 Impact of food waste on national food security

	Frequency	Valid percent	Cumulative percent
Disagree	216	59.1	59.1
Agree	91	24.8	40.9
Neutral	58	16.1	100.0
Total	365	100.0	

Source Author Analysis

Table 3.12 Adverse effect of food waste on social atmosphere

	Frequency	Valid percent	Cumulative percent
Disagree	31	8.6	8.6
Agree	263	72.0	91.4
Neutral	71	19.4	100.0
Total	365	100.0	

Source Author Analysis

respondents) wastes food of an amount (≈) 100 g food at restaurants in each visit, which is supposed be enough to feed approximately twenty-six (26) people once a day. Furthermore, the result of the study bespeaks the willingness of people to adhere to anti-food waste law (83.8%) as they acknowledge their responsibility to save food (82.6%). The fourth section denotes to draft an anti-food waste law respondents' viewpoint toward the law is also examined (Table 3.13).

The results revealed that an anti-food waste system should be developed with an aim to preventing food waste and safeguarding food security by establishing a mechanism specifying responsibilities of various stakeholders (see Table 3.13). The respondents agreed the mechanism might include (a) restaurants to offer customers to order meals divided into large and small portions, and the price will change accordingly; (b) offer a convenient takeaway of leftover food; and (c) creating social awareness about food waste. Furthermore, the respondents (96.7%) also agreed an adequate government policy can reduce food waste (Tables 3.14 and 3.15) which the respondent agreed to follow (83.9%). In order to examine whether food waste at restaurants, a place after house where maximum people consume food, is a usual manner of the restaurant visitors which can contribute in food insecurity, the findings of the web-based survey substantiate regular food waste behavior of the restaurant visitors/

Table 3.13 Framework of anti-food waste law

	Frequency	Valid percent
(A) Law to prevent food waste	255	69.8
(B) Establishes a long-term mechanism to prevent food waste	259	70.9
(C) Aimed at safeguarding food security	259	70.9
(D) Specify the responsibilities of various stakeholders	232	63.5
(E) Restaurants to offer customers to order meals divided into large and small portions, and the price will change accordingly	157	43.0
(F) Restaurants can offer a convenient takeaway of leftover food	114	31.2
(G) Creating awareness about food waste and letting people know that food is hard to come	71	19.4
(H) Fine	24	6.5

Source Author Analysis

Table 3.14 4Need of an adequate government policy to reduce food waste

	Frequency	Valid percent	Cumulative percent
Agree	263	72.0	72.0
Neutral	90	24.7	96.7
Disagree	12	3.3	100.0
Total	365	100.0	

Source Author Analysis

Table 3.15 Wiliness to adhere to anti-food waste law

	Frequency	Valid percent	Cumulative percent
Yes	306	83.9	83.9
No	59	16.1	100.0
Total	365	100.0	

Source Author Analysis

consumers' visiting the restaurants creating enormous quantity of food waste. So, it is important to develop an anti-food waste system.

3.6.3 Attitudes Toward Food Waste Across Different Cultures and Regions

Countries around the world do not share a common ethic on food waste, as attitudes toward food waste vary widely between different cultures and regions. In some countries, food waste is seen as a significant problem and there are concerted efforts to reduce it through education campaigns, regulations, and other measures. In other countries, food waste is not considered a major issue, and there may be less emphasis on reducing it.

In some cultures, wasting food is seen as a moral or ethical failing, as it is seen as disrespectful to the effort and resources that went into producing it. For example, in Japan, the concept of *mottainai* encourages people to use resources carefully and avoid waste. Similarly, in many religious traditions, including Islam and Judaism, there are prohibitions on wasting food, as it is seen as a violation of the principles of hospitality and gratitude. However, in other cultures, wasting food may be seen as a symbol of wealth and abundance, and may be viewed as a desirable status symbol. In these cultures, there may be less social pressure to avoid waste, and more emphasis on indulgence and excess.

Despite these cultural differences, there is growing recognition around the world that food waste is a significant problem, both in terms of its environmental impact and its contribution to global food insecurity. Therefore, many countries are taking steps to reduce food waste through a variety of initiatives, such as reducing portion sizes, improving storage and preservation techniques, and diverting food waste to composting or animal feed, in the European Union, member states are required to implement measures to reduce food waste under the Waste Framework Directive. In the United States, the Food Recovery Hierarchy has been established as a guide for food recovery and waste reduction. Similarly, other countries, like China,[40] France,[41] South Korea,[42] Italy,[43] Australia,[44] and so forth have also adopted legal frameworks and work plan to reduce food waste. So, there is a growing recognition of the need for legal mechanisms to address food waste at the national and international levels with increasing awareness of the social and economic costs of food waste.

3.6.4 Challenges and Opportunities for Enhancing Legal Cooperation and Coordination in Addressing Food Waste

The legal challenges and opportunities for addressing food waste and food insecurity from five perspectives:

1. **Conceptualization and quantification of food waste and food insecurity:** The lack of a clear and consistent way to define and measure food waste and food insecurity across the world impedes the ability to gather and compare data and track progress in different countries and regions. Different legal sources and principles may use different terms and indicators to address food waste and food insecurity, such as food loss, food surplus, food recovery, food donation, etc. Moreover, different parts of the food supply chain may encounter different causes and consequences of food waste and food insecurity, which require different legal measures and solutions. Hence, there is a need to develop a clear and consistent definition and measurement of food waste and food insecurity that can be applied across different levels of governance and jurisdictions, as well as across different stages of the food supply chain. This would enable data collection, comparison, monitoring, and reporting of food waste and food insecurity. Furthermore, this would improve legal clarity, consistency, and accountability for preventing and reducing food waste and ensuring food security.
2. **Harmonization and Coordination of Legal Standards and Policies on Food Waste**: The lack of alignment and integration of legal norms and policies on food waste among different levels of governance and jurisdictions may pose barriers and discrepancies for preventing and reducing food waste along the

[40] [China]Anti-food waste work plan [S] 2022.

[41] [France] LOI n 2016-138 du 11 fvrier 2016 relative la lutte contre le gaspillage alimentaire (1) [S] 2016.

[42] A. D. Steffen, South Korea Has Almost Zero Food Waste, Here's How [EB/OL] www.intelligentliving.co/south-korea-zero-food-waste/ 2019-05-04/2020-03-11.

[43] [Italy] Provisions concerning the donation and distribution of products food and pharmaceuticals for purposes of social solidarity and for the limitation of waste [S] 2016.

[44] [Australia] National Food Waste Strategy [S] 2017.

food supply chain. Different legal sources and principles may employ different rules, categories, labels, reports, monitoring, etc., of food waste. Furthermore, different legal systems may use different methods and tools for regulating and enforcing anti-food waste measures, such as laws, regulations, guidelines, incentives, sanctions, etc. Therefore, there is a need to establish a fair and effective legal responsibility and liability regime for food waste that can assign roles and obligations to different actors and stakeholders along the food supply chain, as well as provide incentives and sanctions for preventing and reducing food waste. This would foster behavioral change, innovation, and collaboration among different actors and stakeholders. Furthermore, this would ensure legal justice, compensation, and deterrence for food waste. Additionally, there is a need to enhance legal cooperation and coordination among different levels of governance and jurisdictions on food waste, such as through harmonizing and aligning legal standards and policies on food waste; facilitating cross-border trade and movement of food products; establishing regional or global platforms or networks for information sharing, dialogue, and cooperation; etc. This would eliminate obstacles and inconsistencies for preventing and reducing food waste along the food supply chain. Furthermore, this would promote legal integration, coherence, and effectiveness for food waste.

3. **Protection and Promotion of Human Rights Related to Food Waste**: The respect and fulfillment of human rights related to food waste and food insecurity, such as the right to adequate food, the right to health, the right to a clean environment, etc., is needed (Chilton and Rose 2009). Different legal sources and principles may interpret and apply these human rights differently in relation to food waste and food insecurity. The human rights obligations and mechanisms related to food waste and food insecurity need to be strengthened, such as by incorporating the right to food and other related human rights into national constitutions, laws, and policies; ensuring the participation and consultation of rights-holders and duty-bearers in the design and implementation of anti-food waste measures; providing access to justice and remedies for victims of food waste and food insecurity; etc. (Godrich et al. 2021; McClain-Nhlapo 2004). This would protect and promote human dignity, well-being, and empowerment in relation to food waste and food insecurity. Furthermore, this would ensure legal recognition, protection, and fulfillment of human rights in relation to food waste and food insecurity.
4. **Enforcement and Compliance of Anti-Food Waste Measures:** The effectiveness and efficiency of anti-food waste enforcement and compliance along the food supply chain is a challenge. Different legal sources and principles may use different mechanisms and instruments for enforcing and monitoring anti-food waste measures, such as inspections, audits, certifications, penalties, etc. (Priefer et al. 2016). Moreover, different legal systems may face different barriers and constraints for ensuring compliance with anti-food waste measures, such as lack of resources, capacity, awareness, incentives, etc. The mechanisms for anti-food waste measures need to be improved, such as by providing adequate resources, capacity, awareness, incentives, etc., for implementing and monitoring anti-food

waste measures; establishing independent and transparent oversight and accountability bodies or mechanisms for anti-food waste measures; developing effective and proportionate penalties or sanctions for non-compliance with anti-food waste measures; etc. (De Laurentiis et al. 2020; Priefer et al. 2016). This would ensure the effective implementation and monitoring of anti-food waste measures along the food supply chain. Furthermore, this would ensure legal enforcement, compliance, and accountability for anti-food waste measures.

5. **Legal Disputes and Conflicts over Food Waste:** Different legal sources and principles may employ different rules and procedures for resolving legal disputes and conflicts over food waste, such as litigation, arbitration, mediation, negotiation, etc. Furthermore, different legal systems may have different benefits and drawbacks for resolving legal disputes and conflicts over food waste, such as access to justice, costs, time, fairness, etc. Legal disputes related to food waste can arise from various issues along the food supply chain. For example, there may be disputes between food producers and retailers over the quality and safety of food products, or between consumers and retailers over the labeling and packaging of food products. There may also be disputes between different levels of governance and jurisdictions over the regulation and enforcement of anti-food waste measures. Furthermore, there may be disputes over the ownership and liability of food waste, such as who is responsible for disposing of food waste or who is liable for damages caused by food waste (Trabold and Nair 2018). For instance, in some countries, food producers are required to pay a fee for disposing of their food waste in landfills or incinerators. There is a need to promote alternative dispute resolution methods (ADR) for resolving conflicts over food waste among different actors and stakeholders along the food supply chain, as well as among different levels of governance and jurisdictions. Such methods may include mediation, arbitration, negotiation, etc. (Hohmann 2018; Işık 2020). This would enable the peaceful and constructive resolution of conflicts over food waste. Furthermore, this would ensure legal dialogue, cooperation, and harmony for food waste.

3.7 Conclusion

To explore whether or not food wastes is exacerbating food insecurity, the study has analyzed the relevant legal sources and principles for addressing food waste and food insecurity at different levels of governance and jurisdictions. It has also discussed the main legal issues and challenges arising from food waste and food insecurity in the global food system. Based on this analysis, this study findings that:

Firstly, food waste and food insecurity are interrelated and complex legal problems that require a holistic and multidisciplinary approach to address them effectively and efficiently.

Secondly, legal sources and principles play a crucial role in preventing and reducing food waste and ensuring food security, as they provide norms, standards, guidance, incentives, sanctions, etc., for different actors and stakeholders along the food supply chain.

Thirdly, legal sources and principles on food waste and food insecurity vary and diverge among different levels of governance and jurisdictions, as well as among different stages of the food supply chain, which may create legal gaps, barriers, inconsistencies, or conflicts for preventing and reducing food waste and ensuring food security.

Fourthly, legal sources and principles on food waste and food insecurity need to be harmonized and coordinated among different levels of governance and jurisdictions, as well as among different stages of the food supply chain, to ensure legal integration, coherence, and effectiveness for preventing and reducing food waste and ensuring food security.

Fifthly, legal sources and principles on food waste and food insecurity need to be strengthened and improved in terms of their definition, measurement, responsibility, liability, enforcement, compliance, dispute resolution, human rights protection, etc., to ensure legal clarity, consistency, accountability, justice, etc., for preventing and reducing food waste and ensuring food security.

Sixthly, the meta-analysis finds 36–51% of food is wasted across the global food supply chain and approximately only 49–64% of food is consumed. Further, developing the "Food Waste Severity Measuring Scale" (Fig. 3.2) and accumulating the findings of the meta-analysis into the measuring scale, the study asserts, with the *iFw* value between 36% and 51% food waste exacerbates food insecurity. Furthermore, by identifying how food waste varies across the global food supply chain, the meta-analysis provided important indications toward policymaking. Additionally, it sheds light on the insufficiency of food waste data across the global food supply chain and identifies data gaps, particularly in food service level.

Addressing the data gap to estimate food waste at food service level, particularly at restaurants, the study conducted web-based survey. The results of the "statistical analysis" denoted that out of 94.6% of respondents visiting restaurants, 77.7% wastes food of an amount ($\approx$) 100 g food at restaurants in each visit, the quantity of food wasted can feed approximately twenty-six (26) people once a day. If 77.7% people generate food wasted of a quantity that can feed approximately twenty-six (26) people once a day, then, it can be deduced that similar quantity of food waste is also generated at school or office canteens, and event venues. Furthermore, the results of the descriptive frequency analysis estimate the rate of people wasting food at restaurants is $\geq$22.5%. So, the findings of the study substantiate food waste exacerbates food insecurity in terms of food quantity. The results of the study denote the need of creating legally binding obligations by setting up a legal and institutional foundation for anti-food waste system to reduce global food waste. Hence, the study recommends a guideline for anti-food waste measures within *the Law of Food Security* (discussed in Chap. 6, Sect. 6.3.8), which aims at reducing food waste across the global food supply chain and advocates for a sustainable utilization method.

References

Aschemann-Witzel J, de Hooge I, Amani P, Bech-Larsen T, Oostindjer M (2015) Consumer-related food waste: causes and potential for action. Sustainability 7(6):6457–6477. https://doi.org/10.3390/su7066457

Blichfeldt BS, Mikkelsen M, Gram M (2015) When it stops being food. Food Cult Soc 18(1):89–105. https://doi.org/10.2752/175174415x14101814953963

Bongaarts J (2021) FAO, IFAD, UNICEF, WFP and WHO. The state of food security and nutrition in the world 2020. Transforming food systems for affordable healthy diets FAO, 2020, 320 p. Popul Dev Rev 47(2):558. https://doi.org/10.1111/padr.12418

Buzby JC, Farah-Wells H, Hyman J (2014) The estimated amount, value, and calories of postharvest food losses at the retail and consumer levels in the United States. SSRN Electron J. https://doi.org/10.2139/ssrn.2501659

Chilton M, Rose D (2009) A rights-based approach to food insecurity in the United States. Am J Public Health 99(7). https://doi.org/10.2105/AJPH.2007.130229

De Laurentiis V, Caldeira C, Sala S (2020) No time to waste: assessing the performance of food waste prevention actions. Resour Conserv Recycl 161. https://doi.org/10.1016/j.resconrec.2020.104946

FAO (2014) Food losses and waste in the context of sustainable food systems A report by The High Level Panel of Experts on Food Security and Nutrition. www.fao.org/cfs/cfs-hlpe [Accessed on 2020-03-11]

FAO (2019) The State of Food and Agriculture 2019. FAO. https://doi.org/10.4060/ca6030en

FAO (2020) The state of food security and nutrition in the world 2020. Food and Agriculture Organization of the United Nations

FAO, I. et al (2021) The state of food security and nutrition in the world 2021. UNICEF Data. https://data.unicef.org/resources/sofi-2021/ [Accessed on 2022-09-21]

Godrich SL, Barbour L, Lindberg R (2021) Problems, policy and politics – perspectives of public health leaders on food insecurity and human rights in Australia. BMC Public Health 21(1). https://doi.org/10.1186/s12889-021-11188-8

Gustavsson J, Cederberg C, Sonesson U, van Otterdijk R, Meybeck A (2011) Global food losses and food waste: extent, causes and prevention. International Congress: Save Food!

Hartikainen H, Riipi I, Katajajuuri J-M, Silvennoinen K (2019) From measurement to management: food waste in the Finnish food chain. In: Food waste management. Springer International Publishing, pp 415–439. https://doi.org/10.1007/978-3-030-20561-4_16

Hohmann B (2018) Possibilities for the application of alternative dispute resolution methods in the administrative procedure. Eur J Multidiscip Stud 3(4). https://doi.org/10.26417/ejms.v3i4.p90-98

Ishikawa T (2014) Reporting on efforts by working group for business practice improvement regarding food waste reduction. Mater Cycles Waste Manag Res 25(1):43–54. https://doi.org/10.3985/mcwmr.25.43

Işık S (2020) Mediation as an alternative dispute resolution method and mediation process in Turkish law system: an overview. Ann Fac Droit Istanb

Katajajuuri J-M, Silvennoinen K, Hartikainen H, Heikkilä L, Reinikainen A (2014) Food waste in the Finnish food chain. J Clean Prod 73:322–329. https://doi.org/10.1016/j.jclepro.2013.12.057

Koivupuro H-K, Hartikainen H, Silvennoinen K, Katajajuuri J-M, Heikintalo N, Reinikainen A, Jalkanen L (2012) Influence of socio-demographical, behavioural and attitudinal factors on the amount of avoidable food waste generated in Finnish households. Int J Consum Stud 36(2):183–191. https://doi.org/10.1111/j.1470-6431.2011.01080.x

Lipinski B, Hanson C, Lomax J, Kitinoja L, Waite R, Searchinger T (2013) Reducing food loss and waste. Installment 2 of creating a sustainable food future. Washington, DC: World Resource Institute, June

Maxwell JA (2022) Interactive approaches to qualitative research design. In: The SAGE handbook of qualitative research design. SAGE, pp 41–54. https://doi.org/10.4135/9781529770278.n4

McClain-Nhlapo C (2004) Implementing a human rights approach to food security, 2020 Africa conference brief, 13

Møller H, Hanssen OJ, Gustavsson J, Östergren K, Stenmarck Å, Dekhtyar P (2014) Report on review of (food) waste reporting methodology and practice. Ostfold Research

Papargyropoulou E, Lozano R, Steinberger JK, Wright N, bin Ujang, Z. (2014) The food waste hierarchy as a framework for the management of food surplus and food waste. J Clean Prod 76:106–115. https://doi.org/10.1016/j.jclepro.2014.04.020

Parfitt J, Barthel M, Macnaughton S (2010) Food waste within food supply chains: quantification and potential for change to 2050. Philos Trans R Soc B Biol Sci 365(1554):3065–3081. https://doi.org/10.1098/rstb.2010.0126

Porpino G, Parente J, Wansink B (2015) Food waste paradox: antecedents of food disposal in low income households. Int J Consum Stud 39(6):619–629. https://doi.org/10.1111/ijcs.12207

Priefer C, Jörissen J, Bräutigam KR (2016) Food waste prevention in Europe – a cause-driven approach to identify the most relevant leverage points for action. Resour Conserv Recycl 109. https://doi.org/10.1016/j.resconrec.2016.03.004

Stancu V, Haugaard P, Lähteenmäki L (2016) Determinants of consumer food waste behaviour: two routes to food waste. Appetite 96:7–17. https://doi.org/10.1016/j.appet.2015.08.025

Thyberg KL, Tonjes DJ, Gurevitch J (2015) Quantification of food waste disposal in the United States: a meta-analysis. Environ Sci Technol 49(24):13946–13953. https://doi.org/10.1021/acs.est.5b03880

Trabold TA, Nair V (2018) Conventional food waste management methods. In: Sustainable food waste-to-energy systems. https://doi.org/10.1016/B978-0-12-811157-4.00003-6

United Nation Environment Programme (2021) Food waste index report 2021. UNEP

Xue L, Liu G (2019) Introduction to global food losses and food waste. In: Saving food. Elsevier, pp 1–31. https://doi.org/10.1016/b978-0-12-815357-4.00001-8

Chapter 4
The Evolution of International Food Law: From the League of Nations to Present Day

4.1 Introduction

International Food Law is an evolving branch of international law whose existence is becoming defined with time since the notions food insecurity, food security, and food safety emerged and was acknowledged in the international arena. Many international food guidelines and agreements emerged but focused on food safety. Consequently, food insecurity remains a persistent global issue. The WEF in its *Global Risk Report 2020* has acknowledged "food insecurity" as a "societal risk" for the next 10 years (WEF 2020). According to reports, since 2014, the number of hungry people worldwide has been gradually rising. If this trend continues, the number will exceed 840 million by 2030 from 690 million (FAO 2020). This means, instead of achieving the Sustainable Development by 2030, zero hungry and good health, we are at the risk to face severe "food insecurity."

To analyze international food law specific notions like food insecurity, food security, food safety, food aid and assistance, right to food, food governance, and food sovereignty need to be focused. Mostly, because the international food law is circulating and evolving around these interconnected notions. The Committee on World Food Security (CFS) in 2009 illustrating the "food security" stated, "food security exists when all people, at all times, have physical, social and economic access to sufficient, *safe and nutritious food* to meet their dietary needs and food preferences for an active and healthy life [...]."[1] Similar to the illustration of "food security" by the CFS, many international organizations, such as FAO, and CODEX, mandated to work in the field of food security and safety, has come forward with guidelines,

[1] CFS. Reform of the Committee on World Food Security Final Version. CFS:2009/2 Rev.2. [S] 2009: 1.

The first draft of this Chapter got published in Asian Journal of WTO & International Health Law and Policy, Vol. 16, No. 2 (SSCI Journal).

S. S. Silvee, *The Law of Food Security: The Legal Framework for a Sustainable and Equitable Global Food System*, LITES - Legal Issues in Transdisciplinary Environmental Studies 5, https://doi.org/10.1007/978-981-95-3636-8_4

policy declarations, and codes of conduct to set standards of conduct. Although these international instruments together contribute in the formation of international food law, but are neither strictly binding in nature nor completely lacks legal significance. Hence, to analyze international food law, scrutinization had to begin from the notion "food insecurity, and food security" along with the international instruments lead by the notions.

Considering the contemporary international instruments focusing on food, promoted by the international organizations, this study aims to explore how the international food instruments emerged; second, its development over time and the contribution of the international organizations in operating the global food system; lastly, can the present international instruments be collectively called as international food law. This chapter analyzes the research question: What are the aim, objectives, nature, scope, and legal significance of contemporary international food law? How the contemporary international food law developed and evolved? How much effective the contemporary international food law is on building food security?

The chapter is divided into three major parts including introduction and conclusion. The discussion begins with a brief history of the international food law portraying how gradually this branch of international law developed and is evolving.

4.2 Source of Development and Evolution of International Food Law

The period following the First World War and preceding the establishment of the United Nations, marked by the existence of the League of Nations, was a critical phase in the international recognition of food insecurity. While the devastating impact of war on food supplies was evident during the Great War, the subsequent two decades saw a nascent but evolving understanding of food scarcity as an issue that transcended national borders and demanded international attention. This era laid the groundwork for future global food security initiatives, even as it grappled with persistent challenges and the horrifying weaponization of hunger. Hence, understanding the history of the contemporary international food law is essential to build a sustainable and equitable global food system. Hence, the study has to begin with analyzing the development and evolution in perception of food security which shaped the contemporary global food system over the 100 years under five periods, from 1920 to 2020.

4.2.1 *Post First World War and League of Nations (1920–1945): Emergence of Food Insecurity in the International Arena*

The notion food insecurity is not an emerging one; it has a long history (Spitz 1985). Addressing food insecurity, Shelton noted in "*The Duty to Assist Famine Victims* "the phenomena of hunger and famine have been recorded throughout history,"[2] which caused increase in malnutrition. For instance, during invasion of the Soviet Union, Nazi German forces deliberately enacted a policy to starve the city of Leningrad (now St. Petersburg) into submission. The German army completely blockaded the city, cutting off all food and fuel supplies. Analyzing the early centuries, it is observed that beside the phenomena of hunger, famine, man-made starvation, and malnutrition, food crimes such as, adulteration, falsification weight specification, and so forth also holds a long history (Shelton 1984). However, countries and international organizations placed a greater emphasis on food crimes rather than the broader issues of hunger, famine, man-made starvation, and malnutrition. Hence, sufficient amount of food safety laws is implemented in early centuries, such as, the Assize of Bread regulation of 1646, the Treatise on Adulterations of Food and Culinary Poisons of 1820, the Victorian Public Food Act of 1854, the UK Adulteration of Food and Drink Act of 1860, the Pure Food and Drug Act of 1906, the Federal Meat Inspection Act of 1906, the Bengal Food Adulteration Act of 1919, and so forth were implemented at different times (Hutt 2001; Hutt and Hutt II 1984; "The Food Safety Information Handbook" 2002; Weber 1974; Wu and Rodricks 2020). So, it can be asserted, the author Karapinar rightly identified that the notion food insecurity mainly developed gradually with the phenomenon of chronic food insecurities, malnutrition, and food safety (Karapinar 2010).

After the First World War the extent to which the food insecurity phenomenon affected millions of people became perceptible, but neglecting the urge of a global food system, the League of Nations perceived the imperative of a global health system to eradicate malnutrition and created the Health Organization of the League of Nations in 1922. Nevertheless, the issue of food insecurity remained as a camouflaged global issue, as it was considered mostly in country or regional levels. But soon the issue of food insecurity became perceptible when the world experienced a global economic crisis in 1929 which had an impact on the then free trade policy and international trade.[3] The global economic crisis made the connection between trade and food insecurity discern when the food-importing countries raised their tariff barriers in the name of protecting domestic industries, leading to a 60% drop in wheat imports, which eventually collapsed trade[4] and increased the price of food; resulting in affecting access to food, affordability of food, and availability of food.

---

[2] Shelton, D. The Duty to Assist Famine Victims [J] IOWA L. REV. 1985, 70 (1309).

[3] Ruoshi, Hou. History shows trade protectionism is short-lived [EB] http://chinaplus.cri.cn/opinion/opedblog/23/20180822/173553.html. 2018-08-22/2020-10-21.

[4] FAO.ORG. FAO 70th Anniversary [EB] http://www.fao.org/70/1945-55/en/ n.d./2020-02-02.

Such phenomenon resulted in legal framework formulated by many countries (van der Meulen 2018). Consequently countries like, United States, India, South Africa, adopted enactments, such as, the Agricultural Adjustment Act (1933), the Federal Food, Drug, and Cosmetics Act (1938), The Bombay Prevention of Food Adulteration Act (1925), the Food, Drugs and Disinfectants Act (1929), Public Health Act (1919) (Stevens 2019). These collide food enactments where implemented aimed to regulate the production, supply, distribution, food safety, and food trade.

In the early 1935, the Health Committee of the Health Organization has submitted a report titled as *Nutrition and Public Health*, the first account of the extent of hunger and malnutrition in the world. The report introduced food insecurity to the international political arena (Russell 1953) According to the report, there was an acute food shortage in the poor countries (Shaw 2007). However, the report couldn't make that impact until, the relationship of food with health, economics, and politics was proposed by Stanley Bruce in the Assembly of the League of Nations. This multidisciplinary aspect of food insecurity roused considerable enthusiasm in the international political arena; as the governments were engaged in resolving political difficulties related to the agricultural sector and economic sector without considering and accepting the benefits of cooperation. Stanley Bruce's multidisciplinary aspect of food insecurity had induced the understanding that to resolve the food insecurity we need to build food security which can be achieved through cooperation because it will give economic and political benefits. So, the governments gave a chance to cooperation for food security, mostly, for economic and political benefits. However, neither the notion food security, nor the notion food insecurity was defined by the Health Committee of the Health Organization, although about 200 definitions in published writings were available in literature (Maxwell and Smith 1992). Consequent of Stanley Bruce's multidisciplinary aspect, the "International Standard of Food Requirements" was agreed, the beginning of coordinated nutrition law, which gave an indication of the amount of food needed throughout the world. A "mixed committee," as a leading authority on nutrition, agriculture, and economics was also created to examine and make recommendations on every aspect of the food problem, including production, transport, and trade (Maidana-Eletti 2014). The committee's first report the Relation of Health, Agriculture and Economic Policy (1937), presented the recent trends in food habits throughout the world, with particular reference to consumption, prices, the relation of income and education to nutrition, and the problem of agricultural adaptation to nutritional requirements, along with the numerous recommendations for future progress (Sassi 2018). It won't be imprecise to assert that during the period of 1920–1945, gradually, food insecurity was considered as a global issue and gradually the nexus with trade, public health, and economy was comprehended.

Hence, in 1938, 22 nations, including the United States and Russia, met in conference to arrange how the world food policy could be carried out to prevent malnutrition and maintain good health (Margulis 2017). However, the outbreak of the Second World War put a hold on this optimistic development. Nevertheless, United Nations

Conference on Food and Agriculture in 1943 with the participation of 44 governments created hopes as the Conference decided on the establishment of a permanent organization in the field of food and agriculture.

4.2.2 Post Second World War and United Nations (1945–1970): Shift Toward Food Security

In the post Second World War, a paradigm shift it noticed. Emphasis was given more to the notion food security than food insecurity along with building of a food aid and food assistance system. The FAO was established as a specialized agency of the UN under *General Assemble (G/A) Resolution 50 (1)* in 1945 aiming to ensure freedom from hunger. Later WHO was created in 1948 "to improved public health conditions, epidemic control, quarantine measures, and drug standardization."[5] Afterwards World Food Programme was also created in 1961 "to achieve Zero Hunger by facilitating the transfer of knowledge, skills, resources and technical know-how, as well as through policy, advocacy and regional collective action, to strengthen national capacities."[6]

FAO has conducted many researches and performing on-the-spot investigations, such as, Uses of Agricultural Surpluses to Finance Economic Development in Under-Developed Countries: A Pilot Study in India Commodity (1995), Study of an Emergency Food Reserve (Resolution 16/51), Report of Working Party on an Emergency Famine Reserve (CCP 52/55), and so forth. After conducting many researches, similar to the "Relation of Health, Agriculture and Economic Policy" prepared by the mixed committee in 1937, FAO concluded by stating "poverty lay at the root cause of hunger and malnutrition."[7] This means economic instability is the core cause of hunger and malnutrition. This gave the notion food security an economic nature, and disseminated a camouflaged message that is—"people are suffering from hunger and malnutrition because they are unable to afford adequate food. And economic development and market stability is solution to address it." The notion received a new dimension, from food as biological necessity to food as a commodity and essence for economic growth. Hence, later emphasis was given in producing surplus food.

In 1954, with the view that surplus commodities can be utilized for the promotion of economic development, special welfare distribution and emergency relief, FAO came up with "*FAO Principles of Surplus Disposal and Guiding Lines* (FAO Principles)." Predominantly, *FAO Principles* was the first guideline that impliedly encouraged international food trade by emboldening its members for "the constructive use of surplus agricultural commodities and at the same time safeguards the interest of

---

[5] WHO.ORG. Who we are. [EB/OL] https://www.who.int n.d./2020-10-16.

[6] Ibid.

[7] FAO.ORG. FAO 70th Anniversary. [EB/OL] http://www.fao.org/70/1945-55/en/ n.d./2020-02-02.

commercial exporters and local producers."[8] Hence, *FAO Principles* stressed on the need of full intergovernmental consultations to ensure the use of surpluses in aid of economic development and with proper safeguards to prevent disturbance of markets or the normal pattern of trade. Although *FAO Principles* encouraged international food trade, but under the General Agreement on Tariffs and Trade (GATT 1947), the members were also given the right to impose import and export restrictions on food stuffs on varies grounds under Article XI.

Furthermore, to deal with the impact of food assistance programs on commercial trade and on agricultural production, and to govern the operations under these Principles, FAO Committee established *Consultative Subcommittee on Surplus Disposal* (CSSD) at the same time.[9] The diversification of the food trade was not only apprehended by the international organizations and manufacturers, governments of different nations had grasped how a strong national food industry can significantly contribute in food security but also in economic growth. Hence, the governments of the conventional food exporter countries along with the other countries have put effort to strengthen national food industry, in terms of production and export, by working on their food law.

By 1967, a number of resolutions of the UN, its specialized agencies, and other intergovernmental bodies urged the negotiation of commodity agreements and five successful international commodity agreements had been concluded for wheat, sugar, coffee, tin and olive oil, such as, International Wheat Agreement (1962), and International Grains Agreement (1967), which formed 10% of world trade in primary products. Analyzing these international commodity agreements, three natures of these agreements are been observed—Multilateral contract agreements, international buffer stock, and Export restriction agreements. The main feature of multilateral contract agreements was that "it contained an obligation on importers or exporters to buy and sell certain guaranteed quantities at stipulated maximum or minimum prices whenever the free-market price reached or exceeded those limits."[10] On the other hand, international buffer stock agreements are "aimed to stabilize prices by an obligation to buy whenever the world price fell below a certain minimum level and to sell when the price rose above a certain maximum."[11] Whereas, export restriction agreements aim to secure some degree of stability of prices by restricting exports at certain degree. However, food export restrictions raise concern as food prices increases exacerbating the issues of food insecurity for people in food-importing countries. Having an in-depth study on the export restrictions it can be stated that food export restrictions impact food prices and exacerbates food insecurity. According to Debucquet and Mamun (2022) and Debucquet et al. (2020) International Food Policy

[8] FAO. Reporting Procedures and Consultative Obligations Under the FAO Principles of Surplus Disposal [R] 2001: p. iii.

[9] At the 410th meeting (October 1995) members agreed to change the acronym of the Subcommittee from "CSD" to "CSSD" to avoid conflict with a sister organization in the UN family.

[10] Shaw, John D. World Food Security: A history since 1945 [M] New York: Palgrave Macmillan, 2007: pp. 6–7.

[11] Ibid.

Research Institute (IFPRI) Food Export restrictions from Covid-19 crises to Ukraine-Russia crisis dataset, countries imposed actual ban on export of 34 food stuffs which impacted 7.4% to 17.1% global trade of food. Hence, the WTO members in the 12th Ministerial Conference agreed "not to impose export prohibitions or restrictions on foodstuffs purchased for non-commercial humanitarian purposes,[12] considering that trade..[…] plays a vital role in improving global food security in all its dimensions and enhancing nutrition."[13]

So, these international agreements focused on mainly trade and market stability to build food security more than peoples' right to adequate food, right to have safe food, right to food, and food standard. Gradually international food trade increased, which seemed to be standing in a good stead to ensure food affordability and availability. However, its volt face was unforeseen. With the gradual increase in international food trade, many other persisting issues relevant with food trade gained concern, such as food crime.

Food crime, effecting food safety, is not a new category of crime that emerged in the period. In the early centuries adulteration, falsification weight specification, additions of poisonous substances like lead and aluminum were considered as violations of food safety, fraud against citizens, and risk to public health (Wilson 2009), resulting into punishment (Grimm 2016). Addressing this old concern to protect the health of the consumer and, at the same time, to ensure fair practices in food trade, FAO jointly with WHO established "*the Codex Alimentarius Commission* (CAC)" in 1963 under the *Resolution No. 12/61* with the vision "to create food safety and quality standards to protect everyone everywhere."[14] Since CAC has started working, it has elaborated a "number of guidelines for developing food safety system,"[15] which required implementation in the law to get the binding effect.

Besides, these strategical, and procedural initiatives human rights aspects of food security was also materialized. The Universal Declaration of Human Rights (UDHR) acknowledged the right to have standard life which included food. But the declaration hindered itself from acknowledging the right to food like it acknowledged the right to education and right to life. Later in 1966, the correlation between eradication of hunger and economic development was promoted by International Covenant on Economic, Social and Cultural Rights (ICESCR), which at present is considered as the international document that acknowledged the "right to food." However, ICESCR addresses the right to adequate food and the right to be free from hunger in terms of methods of production, conservation, and distribution, but does not indulge the right to food.

In the period of 1945–1970, the global community shifted from the notion food insecurity to food security. This shift toward food security particularly shifted

[12] WTO. Ministerial decision on world food programme food purchases exemption from export prohibitions or restrictions WT/MIN(22)/29 WT/L/1140 [S] 2022.

[13] WTO. Ministerial declaration on the emergency response to food insecurity WT/MIN(22)/28 WT/L/1139 [S] 2022.

[14] FAO, Codex Strategic Plan 2020–2025 [S] 2020.

[15] CODEX. General Principles of Food Hygiene. CAC/RCP 1-1969 [S] 1969.

the approach toward economy and trade; making a direct nexus between food and economic development through international trade of food. Furthermore, other notions like right to food, food safety, food standard, and food aid and food assistance evolved in this period. But notions like right to food, food standard, and food safety didn't gain much momentum, whereas, food aid and food assistance grasp attention in the period. Hence, UN General Assembly Resolution 827 (IX) on the "*Establishment of a World Food Reserve was* in this regard, under which FAO was requested to conduct an in-depth study of the then popular concept of a "world food reserve" to determine the feasibility of establishing a world food reserve within the framework of the UN, and the feasibility of such a reserve acting as an institution which could contribute to relieve emergency situations and to counteract excessive price fluctuations.

So, in this period there was an attempt to materialize a legal framework. And the challenges of the period were to build cooperation, harmonization in food standard and policy, maintain food safety and quality standard, and facilitate international food trade to ensure market stability, food availability, accessibility, and affordability, and food aid and assistance service.

4.2.3 World Food Crisis (1970–1990): International Response Toward Food Security, Food Safety, and Food Waste

At the beginning of 1970s the world has witnessed the worst food crisis caused by natural disaster and climate change which created an impact on the food production and food availability. This condition transformed many countries to food importer and many to food exporter. So, the 1970s world food crisis was a shift in paradigm which affected the developing countries more but had an impact on the developed countries as well by creating new trading opportunities. The 1970s world food crisis created much more concern in the international political arena by "demonstrating how unpredictable and fragile the world food security situation was, and how quickly it could change."[16] Consequently, the UN World Food Conference of 1974 was organized, which gave the Committee on World Food Security and appointed World Food Council.

The main purpose of the conference was to endorse the main elements of the proposed world food strategy, (Tudge 1976) and understand how the strategy as a whole and its various parts should be implemented, and how additional resources required for its implementation should be mobilized. The conference concluded with adoption of the resolutions, UN General Assembly Resolution 3348 (XXIX), which stated, "every man, woman and child has the inalienable right to be free

[16] Assessment of the World Food Situation, Present and Future. Document E/CONF. 65/3 [C] United Nations World Food Conference. New York: United Nations, 1974.; See, The World Food Problem. Proposals for National and International Action. Document E/CONF. 65/4. [C] United Nations World Food Conference. New York: United Nations, 1974.

from hunger and malnutrition in order to develop fully and maintain their physical and mental faculties."[17] Moreover, the conference concluded promising to increased food production, especially in the developing countries, improve consumption and distribution of food, and building a system of food security.

One of the outcomes of the conference was a shift in approach toward food security, which created a nexus between economy, trade, eradication of hunger, and poverty. The correlation between these factors disseminated a message that hunger is a global issue caused by economic instability, and economic development is the solution, which can be brought by increasing food production and international food trade. However, increasing economic cooperation and food trade transformed the pattern of food production and consumption worldwide and raised serious concerns regarding food safety and food prices.

Analyzing the outcome of the UN World Food Conference, it is observed that the approach toward food security again shifted and this time it created a nexus between economy, trade, eradication of hunger, and poverty. By creating a correlation between each other and it disseminated a camouflage message—hunger is global issue caused because of economic instability, and economic development is the solution which can be brought by increasing food production and international food trade. Hence, Universal Declaration on the Eradication of Hunger and Malnutrition of 1974 clearly states, "the Food problems must be tackled during the preparation and implementation of national plans and programmes for economic and social development, with emphasis on their humanitarian aspects."[18] Since the conference concluded a number of resolutions and reports have been adopted that addressed, economic cooperation and economic development, coupled with hunger and poverty. Few international documents are mentioned in Appendix A. However, these documents lack legal significance.

Increasing economic cooperation increased food trade which transformed the pattern of food production and consumption worldwide[19] and raised serious concerns regarding food safety and food price. Many researchers, including international organizations, have seen food safety separate from food security. The then, food safety included food hazards, such as chemical contaminants, quality of food, misleading information on food label, and so forth. Hence, a range of guidelines addressing different aspects of food safety has been enacted by FAO and other international organizations, such as General Standard for the Labelling of Prepackaged Foods of 1985, Guidelines on Nutrition Labelling of 1985, and so forth.[20] Later in 1987, when 1986 Chernobyl Disaster caused serious disruptions to food production and trade

[17] UN Report of the World Food Conference. Document E/CONF.65/20. [R] UN 1975.

[18] UNGA, World Food Conference. Universal Declaration on the Eradication of Hunger and Malnutrition. UNGA Resolution 3180 (XXVIII [S] 1974.

[19] Lin, Ching-fu. Scientification of Politics or Politicization of Science: Reassessing the Limits of International Food Safety Lawmaking [J] Columbia Science and Technology Law Review, 2013, 15(1).

[20] For more information regarding the food safety guidelines visit CODEX, Codex Texts: Guidelines [DB/OL]. http://www.fao.org/fao-who-codexalimentarius/codex-texts/guidelines/en/ n.d./ 2020-02-22.

of food products, FAO introduced Safety of food in international trade of 1987. By which FAO recommended a number of approaches to the governments on legislation governing food quality and safety compliance to assure food traded be not accidentally contaminated. These guidelines and recommendations required multiple actions from a variety of stakeholders, specially, government and international organizations to be incorporated into law. But after the UN World Food Conference no further optimistic development was noticed. Moreover, beside the food safety issues which gained momentum in this period, the increase in food trade resulted in an increase in trade barriers, which led to an increase in food prices, making food unaffordable and inaccessible to many. The crisis highlighted the interconnectedness of food waste, food security, and larger systems of food production and distribution. Therefore, it is crucial to develop policies that not only address food security but also consider the economic and social aspects of food production and consumption. The whole phenomenon affected the global food market establishing that "food security exists and will get more prominent with increase in international food trade, but hunger co-exists because of poverty." So, one of the challenges for the period was trade barriers and tariff which resulted in increasing food price making food unaffordable and inaccessible.

4.2.4 The Golden Period of the International Food Law: 1990–2000: Paradigm Shift to Food Safety, Food Standard, and Food Waste

Many researchers address 1990s period as the golden period for the notion food security. Firstly because, by the end of twentieth century many countries like, United States (US),[21] European Union (EU),[22] China,[23] came up with their own food law and implemented standards on food quality for public safety. Whether these national laws compliment the international guidelines and recommendations remains a matter of concern. However, the lack of harmonization in the food standards and tariff barriers at national borders were quite evident. Secondly, till 1995 all the international instruments that addressed food issues where either recommendations, guidelines, or certain political commitments based on customary principles, however, establishment of WTO gave three legally binding international instruments—Agreement on

[21] Corley, M. A Brief History of Food Safety Regulations in the United States [EB/OL] http://www.indusoft.com/blog/2016/09/09/a-brief-history-of-food-safety-regulations-in-the-united-states/2016-09-09/ 2020-02-22, See also, FSIS.USDA.GOV. United States Department of Agriculture Food Safety and Inspection Service History [EB/OL] https://www.fsis.usda.gov/wps/portal/informational/aboutfsis/history n.d./ 2022-10-16.

[22] FOOD.GOV.UK. General food law. [DB] https://www.food.gov.uk/business-guidance/general-food-law n.d./2020-08-02.

[23] Luo, Y., & Wu, G. Food Safety Laws and Regulations [M] In Joseph J. Jen and Junshi Chen (eds.) Food Safety in China: Science, Technology, Management and Regulation. John Wiley, 2017: 345–361.

Agriculture, SPS Agreement, and TBT Agreement, of which SPS and TBT Agreement impliedly added legal significant to FAO and Codex recommendations and guidelines. Consequently, a new branch of international law emerged, referred as *International Food Law*. Thirdly, by establishment of WTO main persisting food security issues, such as availability of food, food safety, and market stability were addressed (Janik 2019; Messenger 2017).

Agreement on Agriculture addressing the problem of market access and market stability, attempts "to establish a fair and market-oriented agricultural trading system."[24] Whereas, addressing the persisting issue of food safety while trade, under the SPS Agreement, the WTO member countries agreed to implement measures to protect human, animal (sanitary measures), and plant health (phytosanitary measures) based on scientific principles, sufficient scientific evidence, and assessment of risk.[25] Furthermore, the WTO members ensured that their sanitary and phytosanitary measures will not be arbitrarily or unjustifiably discriminate in nature. Under this agreement, the WTO members can take measures according to "their appropriate level of protection" but the measures shouldn't create obstruction for food trade. Similarly, under the TBT Agreement, WTO members agreed to implement the measures regarding technical requirements and standards (applied to all commodities), such as labeling, for fair practice in trade.[26] Besides addressing the persisting challenges and issues, this period has made an attempt to build a multilateral food trade governance system with accountability. Hence, besides introducing legal binding and cooperation, dispute settlement system was also introduced. Some of the noteworthy mentioned disputes of the period were—*Australia—Measures Affecting Importation of Salmon (Australia- Salmon),*[27] *European Union (formerly EC)—Measures Concerning Meat and Meat Products (EC-Hormones),*[28] *European Union (formerly EC)—Measures Concerning Meat and Meat Products (Hormones),*[29] *Japan—Measures Affecting Agricultural Products (Japan- Agricultural Products)*[30] *and* so forth, which answered and helped in comprehending obligations, interpretation and implementations of SPS Agreements.

In Australia- *Salmon*[31] case, the Appellate Body explored the question regarding the relation between 2.2 and Article 5.1 and stated, "in the event an SPS measure is not based on a risk assessment, as required in Article 5.1, this measure can be presumed, more generally, not to be based on scientific principles or not to be maintained without sufficient scientific evidence within the meaning of Article 2.2. On

[24] Agreement on Agriculture [S] 1995.

[25] SPS Agreement [S] 1995.

[26] TBT Agreement [S] 1995.

[27] DS18: Australia — Measures Affecting Importation of Salmon [S] 2000.

[28] DS26 European Union (formerly EC)—Measures Concerning Meat and Meat Products (EC-Hormones) [S] 2009.

[29] DS48 European Union (formerly EC)—Measures Concerning Meat and Meat Products (Hormones) [S] 2011.

[30] DS76 Japan—Measures Affecting Agricultural Products [S] 2001.

[31] DS18: Australia—Measures Affecting Importation of Salmon [S] 2000.

that basis, the Appellate Body concluded that a violation of Article 5.1 also implied an inconsistency with Article 2.2."[32] A similar view is observed in *EC-Hormones,*[33] where the Appellate Body stated "Articles 2.2 and 5.1 should constantly be read together..[…] Article 5.1 may be viewed as a specific application of the basic obligations contained in Article 2.2 of the SPS Agreement..[…] Article 2.2 informs Article 5.1: the elements that define the basic obligation set out in Article 2.2 impart meaning to Article 5.1".[34]

Furthermore, in *Japan- Agricultural Products,*[35] shedding light on the elements of the requirement under Article 5.6 not to adopt measures which are more restrictive than required to achieve the Member's appropriate level of protection, the Appellate Body stated, "according to the footnote to Article 5.6, a measure is considered more trade-restrictive than required if there is another SPS measure which: (1) is reasonably available taking into account technical and economic feasibility; (2) achieves the Member's appropriate level of protection; and (3) is significantly less restrictive to trade than the SPS measure contested."[36] The period received 30 consultation requests which showed how with gradual increase in international trade have increased disputes. The disputes emerged in the period have indulged the nexus between global public health and trade of safe food, raising concern regarding food borne diseases[37] or non-communicable diseases transmitted through food[38] and genetically modified food.[39] Although food borne diseases or non-communicable diseases transmitted through food has a long history,[40] but genetically modified food grasp attention gradually after 1980. It is noteworthy to mention that, the emergence of GMOs in agricultural and food production introduced several new and contentious issues before the WTO dispute settlement under SPS Agreement TBT Agreement.

---

[32] Appellate Body Report. DS18: Australia—Measures Affecting Importation of Salmon [S] 2000 See also Panel Report. DS292: European Union (formerly EC)—Measures Affecting the Approval and Marketing of Biotech Products [S] 2001.

[33] DS26 European Union (formerly EC)—Measures Concerning Meat and Meat Products (EC-Hormones) [S] 2009.

[34] Appellate Body Report, DS48—European Union (formerly EC)—Measures Concerning Meat and Meat Products (Hormones) [S] 2011.

[35] DS76 Japan—Measures Affecting Agricultural Products [S] 2001.

[36] Appellate Body Report. DS76 Japan—Measures Affecting Agricultural Products [S] para. 93.

[37] Tauxe, R. Outbreak: Foodborne Illness and the Struggle for Food Safety [J] Emerging Infectious Diseases, 2019, 25(12):pp. 23–42.

[38] "The European cholera epidemics of 1830 and 1847 catalysed the evolution of the earliest multilateral governance of communicable diseases. Thus, the link between international law and communicable diseases is rooted in the mid-nineteenth century, more precisely in 1851, when France convened the first International Sanitary Conference". Aginam, Obijiofor. International law and communicable diseases [J] Bulletin of World Health Organization: Special Theme Global Public Health and International Law, 2002.80 (12).

[39] Anderson, K., & Nielsen, C. GMOs, the SPS Agreement and the WTO [M] In K. Anderson, C. McRae, & D. Wilson (eds.) The Economics of Quarantine and the SPS Agreement. The University of Adelaide Press, 2012: pp. 305–331.

[40] Tauxe, R. Outbreak: Foodborne Illness and the Struggle for Food Safety [J] Emerging Infectious Diseases, 2019, 25(12):pp. 23–42.

Here it is noteworthy to mention, *European Communities—Measures Affecting the Approval and Marketing of Biotech Products,*[41] where the United States preferred the neutral terminology of "biotech products" whereas the EC spoke of genetically modified ("GM") crops and GM foods. This dispute has demonstrated one of the contentious issues that SPS and TBT Agreement failed to address competently. Hence, gradually food safety of genetically modified crops and foods became one of the challenges for the period. Besides, how the "exceptions" indulged in these agreements are used politically and diplomatically in different circumstances to create economic pressure was also evident from these disputes. Moreover, these disputes demonstrated that political and cultural fragmentation collides with deep market integration and trade liberalization when it comes to food safety.[42]

SPS Agreement has correlated public health, regulatory agencies, and the food industry. On the other hand, TBT Agreement addressed the technical aspect of the food in trade. TBT Agreement has indulged and contributed in formatting and modification of labeling on the food with scientific and technical information which is relevant while assessing risks to health or the environment prior to instituting protective standards.[43] The TBT Agreement mainly seeks to minimize the extent to which technical regulations can create "unnecessary obstacles" to international food trade by indulging harmonization in technical regulations.[44]

However, these agreements have also been accused of creating trade barriers that make it difficult for developing countries to export their food products. As a result, many developing countries are forced to dispose of their food products due to non-compliance with these standards, leading to significant food waste. For example, the SPS agreement requires that food products meet specific safety and quality standards, such as the absence of certain chemical residues or microbiological contaminants. These standards can be costly to meet, particularly for small-scale farmers and producers in developing countries who lack the resources to comply with these regulations. As a result, their products are often rejected by importing countries, leading to significant food waste and economic losses. For instance, because of the policies of European supermarkets and their direct suppliers, Kenyan smallholders had to waste around 40% of what they grow for European markets (Stuart 2015) despite the fact that the country holds millions of hungry people.

Similarly, the TBT agreement can also create technical barriers to trade that make it difficult for developing countries to export their food products. For instance, some

---

[41] DS292 EC—Measures Affecting the Approval and Marketing of Biotech Products, WTO Doc WT/DS291, WT/DS292, WT/DS293 [S] 2003.

[42] Lin, Ching-fu. Scientification of Politics or Politicization of Science: Reassessing the Limits of International Food Safety Lawmaking [J] Columbia Science and Technology Law Review, 2013,15(1).

[43] Peel, Acqueline., Nelson, Rebecca and Godden, Lee Commentaries GMO Trade Wars: The submissions in the EC — GMO dispute in the WTO. [J] Melbourne Journal of International Law, 2005 (6).

[44] TBT Agreement [S] 1995.

developed countries may impose strict labeling requirements that developing countries are unable to meet. As a result, food products may be rejected, leading to significant food waste.

Furthermore, two major conferences took place in this period—International Conference on Nutrition in 1992 and World Food Summit in 1996. International Conference on Nutrition conferences took place with the view that "there is enough food for all people and the resources to eliminate hunger and to reduce all forms of malnutrition."[45] Despite addressing "approximately 780 million people in developing countries suffer from chronic food deficiency and more than 2000 million people suffer from micronutrient deficiencies;"[46] ignoring the fact that this is caused because of food insecurity. International Conference on Nutrition conferences clearly denied the food insecurity and food security issues and asserted economic problem, such as poverty, as the rationale.

The conference concluded by adopting of the "World Declaration and Plan of action for Nutrition," which focused to eliminate famine, starvation, and nutritional deficiency. However, nearly two decades after the 1974 World Food Conference, in the World Food Summit the notion food security was defined as—"Food security exists when all people, at all times, have physical and economic access to sufficient safe and nutritious food that meets their dietary needs and food preferences for an active and healthy life."[47] However, poverty was considered as the major cause of food insecurity. The conference concluded with Rome Declaration on World Food Security of 1996 and World Food Summit Plan of Action of 1996, which urged for cooperation among different stakeholders. One of the major developments in the period was affirmation of three associated rights of right to food—"the right of everyone to have access to safe and nutritious food, consistent with the right to adequate food and the fundamental right of everyone to be free from hunger."[48] So, it can be asserted that the period gave the foundation of food as fundamental human rights by affirming the right to food. Hence, the period concluded with an evolving notion "right to food,"[49] but focused mostly on building food trade governance, and food safety by impliedly denying the food insecurity and claiming food security exists. The period has clearly demonstrated the perplexing understanding of food security and the establishment of legal frame work based on economic and trade perspective so that the world's trade regime is geared toward building of food security.

---

[45] ICN was proposed at the 15th session of the Subcommittee on Nutrition of the UN Administrative Committee on Coordination (ACC/SCN) in 1989.

[46] FAO Conference in November 1989 and by the World Health Assembly in Mary 1990.

[47] FAO, An Introduction to the Basic Concepts of Food Security [EB/OL] http://www.fao.org/3/al936e/al936e00.pdf n.d./2020-02-10.

[48] Rome Declaration on World Food Security [S] 1996.

[49] World Food Summit Plan of Action [S] 1996.

4.2.5 Persisting Challenges and Right to Food: 2000 to Present

The dawn of the twenty-first century saw the international community grappling with persistent challenges to food security, largely framed within the legal and regulatory architecture of global trade. The World Trade Organization's dispute settlement body became a key arena, resolving numerous conflicts over SPS and TBT agreements. Since the dispute under SPS and TBT gained momentum WTO dispute settlement has resolved many disputes. Such as, *Certain mandatory country of origin labeling (COOL) requirements,*[50] *Certain Country of Origin Labelling Requirements,*[51] *Import Bans, and Testing and Certification Requirements for Radionuclides.*[52] With the increase in international food trade and implementation of protective measures based on SPS and TBT Agreement persisting issues such as, non-communicable diseases (NCDs), nutrition labeling, genetically modified food (GM), and so forth got new dimensions. Since WTO has established it has received 104 cases regarding SPS and TBT Agreement. Of which, 55 involves TBT Agreement[53] and 49 involves SPS Agreement[54] of which 22 involves both the Agreements.

Analyzing the past and present disputes before the WTO dispute settlement body suggests that the disputes circulated mostly regarding the appropriate level of protection measures and food safety issues, such as, food borne diseases, transmit of non-communicable diseases through food, nutrients labeling, food packaging, and so forth. However, one of the major concerns was and is regarding the genetically modified crops and food as there are no international standards for genetically modified crops and food when the modern agriculture is depended on genetically modification to increase food production. However, the Codex Committee on Food Labelling is currently considering the adoption of an international standard on GMO labelling. So, it can be asserted that 2000 year begun with unresolved persisting food issues and challenges. While these trade-related debates over food safety, nutrition labeling, and non-communicable diseases represent significant and ongoing challenges, a more brutal and fundamental threat to the right to food has simultaneously intensified: the deliberate use of starvation as a weapon of war. This tactic fundamentally shifts the challenge from one of regulatory harmonization to the prevention of systemic, man-made famine.

Critically, the post-2000 era has also made visible a dimension that trade law does not directly address but which is central to the right to food: the weaponization of hunger in armed conflict. International humanitarian law prohibits starvation of civilians as a method of warfare, yet contemporary conflicts repeatedly degrade or

---

[50] DS384: United States—Certain Country of Origin Labelling (COOL) Requirements [S] 2015.

[51] DS386: United States—Certain Country of Origin Labelling Requirements [S] 2015.

[52] DS495: Korea—Import Bans, and Testing and Certification Requirements for Radionuclides [S] 2019.

[53] WTO.ORG, Disputes by Agreement [DB/OL]. https://www.wto.org/english/tratop_e/dispu_e/dispu_agreements_index_e.htm n.d./2020-02-25.

[54] Ibid.

deny access to food systems by targeting farms, markets, storage, water, and power infrastructure, and by obstructing humanitarian relief. War acts as a catastrophic accelerant for food crises, transforming localized insecurity into systemic famine by systematically dismantling the very pillars of sustenance: production, distribution, and access. It is not merely a byproduct of conflict but often a deliberate strategy, a weapon that targets the civilian population far from the front lines. The raw mechanics of this process involve the direct destruction of agricultural infrastructure such as farms, irrigation systems, and grain silos; the severing of vital supply chains through blockades, sieges, and the mining of roads; and the collapse of local economies, which renders any available food unaffordable for a populace stripped of their livelihoods. This weaponization of hunger creates a vicious cycle where malnutrition weakens a society's resilience, fueling further instability and prolonging the conflict that created the crisis in the first place.

This destructive dynamic is starkly and tragically illustrated by the situation in Gaza, which serves as a contemporary textbook case of conflict-driven famine. The combination of intense military bombardment and a near-total siege has precipitated a catastrophic collapse of the food system. Agricultural lands have been razed, fishing fleets destroyed, and essential infrastructure like bakeries and water desalination plants have been incapacitated. The severe restriction on the entry of humanitarian aid, including food, water, and fuel necessary for its distribution, has choked off the last remaining lifelines. This has resulted in what international bodies have described as famine-like conditions, with widespread acute malnutrition, particularly among children, and a population forced into a desperate search for sustenance. The situation in Gaza demonstrates with brutal clarity how a targeted and sustained military campaign can, in a matter of months, engineer a full-scale humanitarian disaster by systematically eliminating every means of food self-sufficiency and external support.

The tragedy in Gaza, however, is not an isolated phenomenon but rather the latest chapter in a grimly familiar playbook of modern warfare. In Siege warfare in Syria Government and allied forces, as well as some opposition groups, used "kneel or starve" tactics—encirclement, checkpoint restrictions, and obstruction of aid—leading to acute malnutrition and forced population transfers.

In Yemen, years of conflict and a crippling blockade imposed by the Saudi-led coalition on key ports like Hodeidah systematically starved the nation, leading to one of the world's worst humanitarian crises and leaving millions on the brink of famine. Similarly, the conflict in Sudan has seen warring factions deliberately loot humanitarian warehouses, destroy markets, and turn the nation's agricultural heartland into a battleground, directly causing acute food shortages across the country. In Ethiopia's Tigray region, government forces and their allies were accused of employing a strategy of starvation, systematically destroying harvests, killing livestock, and imposing a blockade that prevented aid from reaching millions. In each instance, the narrative is tragically consistent: food is not just collateral damage; its denial is a core component of military strategy, used to subjugate populations and achieve political ends.

Ultimately, the long-term consequences of war-induced food insecurity extend far beyond the immediate suffering of starvation. They inflict deep, multi-generational scars on a society. Widespread childhood malnutrition leads to irreversible cognitive and physical stunting, crippling a nation's future human capital. The destruction of agricultural capacity, the loss of traditional farming knowledge as communities are displaced, and the contamination of land create a state of protracted dependency on international aid, which itself is often insufficient and politically precarious. Rebuilding a shattered food system is a monumental task that requires not just the cessation of hostilities but also immense investment, political stability, and a concerted effort to restore market functions and ecological health. Therefore, addressing food insecurity in conflict zones cannot be separated from the pursuit of peace and justice; it requires holding perpetrators accountable for using starvation as a weapon of war and recognizing that a sustainable peace can only be built upon the foundation of a fed and resilient population.

The appointment of UN Special Rapporteur on the right to food under resolution 2000/10 added new hopes to resolve food security issues in 2000, however, fall short in addressing the challenges. UN Special Rapporteur was mandated to promote, protect, and identify emerging issues related to the right to food. However, since its inception, the effectiveness of the Special Rapporteur in resolving global food security issues has been a subject of debate. Since its inception, the Special Rapporteur has played a crucial role in highlighting the normative framework of the right to food, advocating for its implementation, and documenting violations. However, despite these efforts, global food insecurity remains a persistent and often worsening challenge.

Establishment of different international organizations and authoritarian has created one of the prominent challenges for the present era of the contemporary global food system—to distinguish between which UN agencies or international agency holds the authoritative power and complementary. In other words, as because of the various channels through which food security can be achieved the policies and functions of the various international organizations overlaps with each other making it difficult to draw the distinction between the main international organization and the relevant international organizations working as complementary. From the above historical analysis, it can be asserted the main UN organ holding the authoritative power has to be UN General Assembly, as it considers Agriculture development, food security and nutrition as an item on its agenda, and Economic and Social Council (ECOSOC), as it considers various reports of committees, funds and programmes that deal with the issue of food security and agriculture. In terms of main international organization, it has to be FAO as it objectives is to help eliminate hunger, food insecurity, and malnutrition, and other organizations such CODEX, WFP, IFAD, UNEP, WHO, WTO are international organization working as complementary. But as one of the international food instruments, i.e. SPS and Agriculture Agreement, enforced by WTO holds legal significance, hence, this often makes the complementary work of WTO as the superintend work. Although WTO as an international organization explicitly recognizes food security with the thought that trade can improve food availability, food accessibility, market stability, and food affordability, it does not make the organization as the principle organization for building food security. However, any international organization can contribute in building food security, as it bestows the right to food that overlaps with economic development, social development, social security, and so forth, and can be built through legal entitlements and secured through accountability mechanisms.

There are a number of international instruments that provided the legal basis for the right to food, such as, article 25 of UDHR, article 11 of the ICESCR, 6 of the ICCPR, however, the right was not defined nor explored under these international instruments. UN Special Rapporteur UNESC (E/CN.4/2001/53) defined the right to food in its first report by stating—"The right to food is the right to have regular, permanent and unobstructed access, either directly or by means of financial purchases, to quantitatively and qualitatively adequate and sufficient food corresponding to the cultural traditions of the people to which the consumer belongs, and which ensures a physical and mental, individual and collective, fulfilling and dignified life free from anxiety." This definition of right to food clearly corollaries that the right to food is food security. Furthermore, the report stated that, the concept of the right to food is made up of different components, such as the right to adequate food, the notions of adequacy and sustainability, dietary needs, cultural or consumer acceptability, and economic accessibility.

Addressing similar components of the right to food in 2009, CFS endorsed a new perspective toward food security. It advocated only access to food alone cannot improve food security, to "improving food security and safety requires addressing some host of issues. For food security, four constituents are needed to be assured – food availability, access, utilization, and market stability."[55] Although, CFC refined from stating that right to food is food security, but it advocated economic approach toward food security based on international trade to ensure the availability of food, food accessibility, market stability, and food affordability. This approach has increased agricultural production creating severe environmental impacts. So the shifting toward Green Technology is also among the challenges. The contemporary global food system focusing on increasing producing and international food trade have shaped to maximize efficiency gains and produce large volumes of commodities, but failed to take distributional concerns into account. Considering all these unresolved issues within the contemporary global food system Olivier De Schutter, UN Special Rapporteur on the Right to Food, "called for the world's food system to be radically and democratically redesigned."[56] However, as these instruments did not fully define or explore the right in detail, its principles are vulnerable. The challenge of the present era is therefore twofold: to resolve the intricate legal questions of global food trade and safety, while simultaneously confronting and holding accountable those state and non-state actors who violate the right to food in its most egregious form, by actively wielding hunger as a weapon of war.

---

[55] Session 12_Committee on World Food Security (CFS Brief). [R] Rome: CFS, 2015 See also, Anderson, K. Agricultural trade, policy reforms, and global food security [M] London: Palgrave Macmillan, 2018: p.4.

[56] UNGA, Report of the Special Rapporteur on the right to food, Olivier De Schutter A/HRC/25/57 [S] 2014.

So, the period from 2000 to the present has been marked by: unresolved regulatory disputes over safety and biotechnology under SPS/TBT; governance overlaps among international institutions; and, increasingly, the use of hunger as a weapon in armed conflicts. Advancing the right to food requires bridging these domains: aligning trade rules with human rights due diligence; safeguarding food systems and humanitarian access during conflict; and strengthening accountability so that no actor can instrumentalize hunger without legal consequence.

4.3 International Food Law Toward Food Security: Historical Perspective and the Six Components of the Global Food System

The approach of international food law toward food security in the global food system has evolved over time, reflecting changes in the way we think about and address food insecurity. Historically, the focus was primarily on hunger, malnutrition, and poverty, and addressing these issues was seen as the key to achieving food security. However, this approach has changed in recent years, as food security has come to be understood as a more complex concept that encompasses a wide range of issues related to food access, availability, quality, and safety. Contemporary international food law is drafted based on three major notions: food insecurity, food security, and food safety. While these instruments focus on a range of issues related to food security, they tend to place greater emphasis on food safety, facilitating food trade, harmonizing food standards, and ensuring the right to safe food. As a result, the contemporary global food system is dependent on a scattered network of binding and non-binding instruments. Despite the proliferation of international food law, there are still significant gaps in our ability to address food insecurity and ensure food security for all. To effectively address these issues, it is necessary to adopt a more holistic approach that takes into account the many factors that contribute to food insecurity, including poverty, inequality, climate change, and environmental degradation. Only by adopting such an approach can we hope to build a truly sustainable and equitable global food system that meets the needs of all people.

The rapid growth of international food trade triggered all aspects of food security advocated by CFS. Besides, it also demonstrated that the market reality was different toward the developing countries. Many developing countries with the hope that international food trade will enhance their economy and resolve food insecurity are still struggling to produce surplus food, to maintain the appropriate level of food standards, to find an appropriate market for business, and to maintain low waste of food in the food production stage. Many developing countries like Bangladesh, China, India, Sudan, Somalia, and so forth, are still struggling to maintain their national food security system which urges for technological and economical support beside cooperation. Analyzing the contemporary issues and law it can be asserted that the contemporary international food law falls-short in addressing food production, food

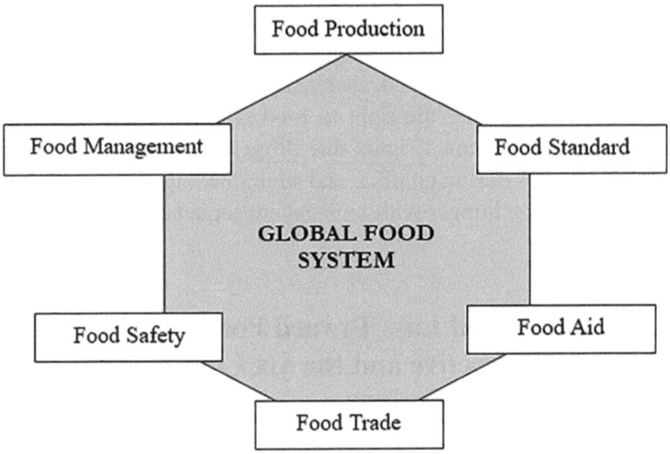

Fig. 4.1 Components of global food system. (Source: Author Analysis)

technology, and food supply chain issues which is entangled with food security. So, the narrow understanding of food security is quite visible to establish a food system which doesn't address all the six (6) major components of the global food system (Fig. 4.1): (1) Food Production, (2) Food Management; (3) Food Standard; (4) Food Safety; (5) Food Trade; and (6) Food Aid.

Here it is noteworthy to mention, even after so many years of emergence of the notion food security and food insecurity, the notion is contemplated from economic perspective, i.e. trade and development, and public health perspective, i.e. food safety, hunger, and malnutrition. The contemporary understanding of food security circulates within the parameter of food safety, hunger, malnutrition, and poverty. However, the contemporary international food law focuses more on food safety during international food trade, which eventually contributed in increased food price. Although, historical analysis of the contemporary international food law suggests the foundation of the contemporary global food system is based on three major concepts food insecurity, food security, and food safety. But the shift toward the concept food safety, hunger, malnutrition, and poverty has impacted the global food system. Hence, despite many international trade cooperation between developed and developing countries, food insecurity, hunger, malnutrition, food price still remains an unresolved issue impacting food security.

Although the *United Nations Development Programme* (UNDP) claims, "the number of undernourished people has dropped by almost half in the past two decades because of rapid economic growth and increased agricultural productivity."[57] Here "a distinction should be drawn between two concepts: hunger or undernourishment,

[57] UNDP.ORG. Goal 2: Zero hunger. [EB/OL] https://www.undp.org/content/undp/en/home/sustainable-development-goals/goal-2-zero-hunger.html n.d./2020-02-2.

and malnutrition."[58] The UN Special Rapporteur report defined hunger or undernourishment as insufficient supply or, at worst, a complete lack of calories. Malnutrition, on the other hand, as by the lack or shortage, in food which otherwise provides sufficient calories, of micronutrients—chiefly vitamins (organic molecules) and minerals (inorganic molecules)."[59] Analyzing this definition of hunger or undernourishment and malnutrition with food security demonstrates a complicated scenario. That is, if "food security exists when all people, at all times, have physical, social and economic access to sufficient, *safe and nutritious food* to meet their dietary needs and food preference," then food security subsists when there is zero, or even minimum, hunger or undernourishment and malnutrition exists. Similarly, zero, or even minimum, hunger or undernourishment subsists when there is sufficient supply of calories which is received from food. So, in other words, if there is sufficient supply of food there will be zero or even minimum, hunger or undernourishment. Analogously, zero, or even minimum, malnutrition exists when the supplied food contains sufficient calories and micronutrients. So, existence of hunger or undernourishment and malnutrition directly questions on the existence of food security by questioning on food production, food supply, and quality of food and it asserts that food insecurity exists.

Accordingly, existence of hunger or undernourishment and malnutrition demonstrates that the claim made in International Conference on Nutrition conferences that "there is enough food for all people and the resources to eliminate hunger and to reduce all forms of malnutrition"[60] is not sufficiently correct. Hence, regrettably, hunger and malnutrition, still, remains as a global issue. Moreover, it is predicted that by 2050 the world pollution will reach to 9.8 billion nearly, which means more than 3 billion more mouths to feed. On the other side there is approximately 56% food gap exists between the crop production and what is need by 2050.[61] Whereas, approximately 1.3 billion tons of food being wasted each year according to FAO reports.[62] This clearly demonstrates that even after 100 years, food insecurity still remains a global issue and will remain if adequate institutionalization, food governance and operating global food system is not build through comprehensive legal framework.

In this trade globalization era, it is difficult to believe that the problem of food insecurity is still wrangling (Nilsson 2004), where the food produced in one-place travels across the borders to another and affect the health and diet of the people living there;

---

[58] UNESC. Report by the Special Rapporteur on the right to food UNESC, E/CN.4/2001/53 [S] 2001.

[59] UNGA. Right to food A/74/164. GA Resolution 73/171 [S] 2019.

[60] ICN was proposed at the 15th session of the Subcommittee on Nutrition of the UN Administrative Committee on Coordination (ACC/SCN) in 1989.

[61] Ranganathanet J et al. How to Sustainably Feed 10 Billion People by 2050, in 21 Charts [EB/OL] https://www.wri.org/insights/how-sustainably-feed-10-billion-people-2050-21-charts 2018/2020-05-21.

[62] Chainey, Ross. Which countries waste the most food? [EB/OL] 2015. World Economic Forum. https://www.weforum.org/agenda/2015/08/which-countries-waste-the-most-food/ 2015/2020-10-08.

(Nicolopoulou-Stamati et al. 2016) raising more concern regarding food safety,[63] as many people and children die (Suggestions on Strengthening Food Safety in the PRC 2006) each year from food borne diseases caused by microbiological agents in contaminated food (Tauxe 2019) and water.[64] Moreover, it is an era where advance technologies are being used at its full-fledged in almost every sector, including, agriculture.[65] This era with technological innovations and new production methods of the industrial revolution have made the food systems (Magomedov et al. 2020) complex and raised concern regarding public health. Nevertheless, food safety and security are a predominate issue. Addressing these concerns requires multiple actions from a variety of stakeholders, such as, government, international organizations, manufactures, and distributor to work in cooperation to create food availability, safety, and quality standards to protect everyone everywhere.[66]

Furthermore, as we delve into the intricacies of international food law, it is crucial to contextualize our study by tracing its evolution over time. The history of food law is a testament to how the global food system has changed and evolved, shaping the current state of affairs. From early civilization's basic regulations on food safety and quality to today's complex international regulatory frameworks, food law has come a long way. Understanding the historical context of international food law provides a framework to appreciate the current challenges and opportunities and predict the future trajectory of food law.

4.4 Conclusion

The contemporary international food law, which is a collection of international instruments regarding food, emerged with the concept "food security." Even after 100 years since the concept "food security" received global acknowledge, the contemporary international food law has failed to resolve "food security" issues (Salvatici 2015). Still there are 690 million people waiting for food which is predicted to increase in near future. Understanding the history of the contemporary international food law is key to build a sustainable and equitable global food system. Analyzing the 100 years of "food security," the present article observed the following –.

---

[63] WTO. Food Safety, Healthy Diets and Trade. [G/OL] In: International Forum on Food Safety and Trade (Conference)GENEVA:FAO/WHO/AU.2019. https://www.wto.org/english/tratop_e/sps_e/01_a_concept_note_food_safety_healthy_diets_and_trade_en.pdf. 2020-02-08.

[64] Lin, Ching-fu. Scientification of Politics or Politicization of Science: Reassessing the Limits of International Food Safety Lawmaking [J] Columbia Science and Technology Law Review, 2013,15(1).

[65] King, A. Technology: The Future of Agriculture [J] NATURE, 2017, Vol. 544 (7651): pp. S21–S23.

[66] FAO, Codex Strategic Plan 2020–2025 [R] Rome: FAO, 2020.

Firstly, the contemporary international food law is drafted based on three major notions food insecurity, food security, and food safety. A change in trend in comprehending the notion food security is also observed which shifted toward hunger or undernourishment, malnutrition, poverty, and safety.

Secondly, the contemporary international food law focuses on food safety, facilitating food trade, harmonizing food standards, and ensuring food rights. These international food law forming the contemporary global food system are depended on scattered binding and non-binding instruments which focuses on food safety. This means international food law emphasizes on food security in terms of consumption by attempting to ensure the food consumed is safe. Hence, international cooperation in the form of a binding treaty or agreement is necessary for an effective response to food security which emphasizes on building food security both in terms of availability, accessibility, and affordability of sufficient quantity of food and consumption of safe.

Thirdly, the contemporary international food law lacks legal binding effect, consequently, has failed to build food governance, address food security issues, general principles, and so on. The only two contemporary international food law having legal significance SPS and TBT agreement, however, SPS mainly focuses on food trade and food safety. So, the four major international food law identified are Codex Alimentarius, SPS Agreement, TBT Agreement, and UN General Assembly Resolution.

Fourthly, the contemporary international food law has failed to bring uniformity in food standard, removing discriminative measures in food trade, and resolving food security and safety issue as it falls short in addressing all the components of the global food system (Fig. 4.1). Fifthly, the contemporary international food law of food safety is enacted from economic aspects. However, the global food system has a humanitarian, ethical, environmental, public health, and social aspect which can only be addressed adequately by a comprehensive approach.

Additionally, analyzing how the international food law emerged, its development over time and the contribution of the international organizations in its development, it is clear that international food law is one of the branches of international which requires multidimensional, cooperative, and comprehensive approach. Food security issues can be resolved only if a global food system comes into existence to function by enacting a comprehensive international food law with just and fair regulations. A comprehensive international food law will give a codified code of fair practice, rights and obligations, guidance for trade and management by creating an institutional framework and food governance, which are prerequisite for a global food system to function efficiently.

This chapter provides an in-depth analysis of the current state of international food law, its regulatory framework, and the challenges it faces. However, to truly understand the complexities of this area of law, it is necessary to examine its history and evolution. The history of international food law is a story of the gradual recognition of the importance of food safety and security, the development of international standards, and the establishment of a framework for cooperation between countries. This evolution has been driven by a range of factors, including major food safety incidents, changes in global trade patterns, and advancements in science and technology. By exploring the history and evolution of international food law, we can gain a deeper understanding of the context in which current laws and regulations have been developed and the challenges they seek to address.

References

Debucquet DL, Mamun A (2022) Documentation for food and fertilizers export restriction tracker: tracking export policy responses affecting global food markets during crisis. Int Food Policy Res Inst. https://doi.org/10.2499/p15738coll2.135857

Debucquet DL, Mamun A, Parent M (2020) Documentation for the COVID-19 food trade policy tracker: tracking government responses affecting global food markets during the COVID-19 crisis. Int Food Policy Res Inst. https://doi.org/10.2499/p15738coll2.133711

FAO (2020) The state of food security and nutrition in the world 2020. Food and Agriculture Organization of the United Nations

Grimm H-U (2016) Information-domination in the European food industry: focus on Germany. In: International food law and policy. Springer International Publishing, pp 3–30. https://doi.org/10.1007/978-3-319-07542-6_1

Hutt PB (2001) Regulation of food additives in the United States. In: Food additives. CRC Press. https://doi.org/10.1201/9780824741709.ch8

Hutt PB, Hutt PB II (1984) A history of government regulation of adulteration and misbranding of food. Food Drug Cosm L J 39(1)

Janik R (2019) Matthias Herdegen, Principles of International Economic Law. Oxford University Press, Oxford et al., 2016, xxxix 624 pp., ISBN 9780198790563, EUR 57.00/Mitsuo Matsushita, Thomas J Schoenbaum, Petros C Mavroidis and Michael Hahn, The World Trade Organization. Law, Practice, and Policy. Oxford University Press, Oxford et al., 2015, lxxxix 944 pp., ISBN 9780199571857, EUR 112.84. Austrian Rev Int Eur Law Online 21(1):431–433. https://doi.org/10.1163/15736512-02101012

Karapinar B (2010) Introduction: food crises and the WTO. In: Food crises and the WTO. Cambridge University Press, pp 1–22. https://doi.org/10.1017/cbo9780511712005.002

Magomedov IA, Khaliev MSU, Ibragimova LV (2020) The need for introducing new technology in agriculture to ensure a sustainable future. IOP Conf Ser Earth Environ Sci 548(3). https://doi.org/10.1088/1755-1315/548/3/032026

Maidana-Eletti M (2014) International food standards and WTO law. Deakin L Rev 19(2):217. https://doi.org/10.21153/dlr2014vol19no2art435

Margulis ME (2017) The forgotten history of food security in multilateral trade negotiations. World Trade Rev 16(1). https://doi.org/10.1017/S1474745616000410

Maxwell S, Smith M (1992) Household food security: a conceptual review. In: Household food security: concepts, indicator, measurement. UNICEF/IFAD, New York/Rome

Messenger G (2017) The World Trade Organization: law, practice, and policy. By Mitsuo Matsushita, Thomas J. Schoenbaum, Petros C. Mavroidis, and Michael Hahn. J Int Econ Law. https://doi.org/10.1093/jiel/jgw074

Nicolopoulou-Stamati P, Maipas S, Kotampasi C, Stamatis P, Hens L (2016) Chemical pesticides and human health: the urgent need for a new concept in agriculture. Front Public Health 4. https://doi.org/10.3389/fpubh.2016.00148

Nilsson H (2004) Food for all – the need for a new agriculture. J Clean Prod 12(5). https://doi.org/10.1016/s0959-6526(03)00107-0

Russell EJ (1953) The white man's dilemma: food and the future. Int Aff 29(4). https://doi.org/10.2307/2606038

Salvatici S (2015) Writing the history of food security since 1945. Contemporanea 18(3). https://doi.org/10.1409/80319

Sassi M (2018) The history of food security: approaches and policies. In: Understanding food insecurity. https://doi.org/10.1007/978-3-319-70362-6_5

Shaw DJ (2007) World food security: a history since 1945. Palgrave Macmillan. https://doi.org/10.1057/9780230589780

Shelton D (1984) The duty to assist famine victims. Iowa L Rev 70. https://heinonline.org/HOL/Page?handle=hein.journals/ilr70&id=1335&div=&collection=/2020-10-16

Spitz P (1985) The right to food in historical perspective. Food Policy 10(4):306–316. https://doi.org/10.1016/0306-9192(85)90022-3

Stevens SK (2019) Tracing the food safety laws and regulations governing traceability: a brief history of food safety and traceability regulation. In: Food traceability. https://doi.org/10.1007/978-3-030-10902-8_2

Stuart T (2015) Food: how much does the world need? World Economic Forum (WEF), May 7. https://www.weforum.org/agenda/2015/05/food-how-much-does-the-world-need/ [Accessed on 2019-12-09]

Suggestions on strengthening food safety in the PRC (2006). https://doi.org/10.22/JS/JQUERY.DATATABLES.MIN.JS

Tauxe R (2019) Outbreak: foodborne illness and the struggle for food safety. Emerg Infect Dis 25(12):2342. https://doi.org/10.3201/eid2512.191192

The food safety information handbook (2002) Choice Reviews Online 39(11). https://doi.org/10.5860/choice.39-6461

Tudge C (1976) Hunger, politics and markets the real issues in the food crisis. Food Policy 1(3):262. https://doi.org/10.1016/0306-9192(76)90061-0

van der Meulen BMJ (2018) Codex Alimentarius: the impact of the joint FAO/WHO food standards programme on EU food law. SSRN Electron J. https://doi.org/10.2139/ssrn.3192451

Weber GA (1974) The food, drug, and insecticide administration: its history, activities and organization, vol 50. Johns Hopkins Press

WEF (2020) The global risks report 2020 insight report, 15th edn. Weforum.Org

Wilson B (2009) Swindled: from poison sweets to counterfeit coffee – the dark history of the food cheats. Princeton University Press. https://doi.org/10.1515/9780691214085

Wu F, Rodricks JV (2020) Forty years of food safety risk assessment: a history and analysis. Risk Anal 40. https://doi.org/10.1111/risa.13624

Chapter 5
Why the Contemporary International Food Law Failed to Ensure Food Security

5.1 Introduction

According to reports by the UN food agency, the number of hungry people increased from 811 million in 2018 to over 820 million in 2020, and 663 million people globally are undernourished (FAO 2020). Analyzing the 100 years since the emergence of the concept "food security" in Chap. 4 and the development of numbers of international guidelines and enactments, it is comprehended that "food insecurity" and related issues are still persisting global issue. This raises question of why, despite the development of numerous international guidelines and enactments over the past 100 years, "food security" still remains a "societal risk." The UN agency reports not only indicate the exacerbating situation of food insecurity but also present the foremost failure of the contemporary legal mechanisms and institutional framework governing the contemporary global food system to uphold the principles of food security. The chapter aims at exploring whether the contemporary international food law aims at building food security.

The contemporary international food law governing the contemporary global food system is examined in this chapter, according to two main areas: firstly, the study explores the rationales behind the exacerbating situation of food insecurity are explored by using Theorizing Regulatory Intermediaries (RIT) model; (Abbott 2017) and secondly, Institutional Analysis and Development (IAD) framework (McGinnis 2011) is used to understand the institutional framework and food governance within the contemporary global food system. The study affirms that the contemporary global food system is complex and requires a more robust institutional framework and governance due to multiple international instruments, which are neither strictly binding in nature nor entirely lacking in legal significance. Additionally, these instruments tend to focus on food safety during food trade, rather than building food security.

Part of this Chapter got published in Asian Journal of WTO & International Health Law and Policy, 2023, 18 (2) (SSCI Journal).

S. S. Silvee, *The Law of Food Security: The Legal Framework for a Sustainable and Equitable Global Food System*, LITES - Legal Issues in Transdisciplinary Environmental Studies 5, https://doi.org/10.1007/978-981-95-3636-8_5

Furthermore, the contemporary international food law needs to establish an operating global food system that can be acknowledged as international food law. Hence, the contemporary global food system is clandestinely exacerbating food insecurity instead of building food security.

Through the historical analysis of the development of the contemporary international food law and the contemporary global food system in Chap. 4, this chapter examines the research question three categories of international instruments relevant to food security: food safety, standards, and rights. Specifically, this chapter analyzes (1) Codex Alimentarius; (2) SPS and TBT Agreement; and (3) UN General Assembly Resolutions. This chapter examines the research question: Do the contemporary international food instrument address all aspects of food security, in terms of the four pillars of food security, food nutritional and safety standard, food waste, risk management, and operating green food system? Do the contemporary international food law protects food rights such as peoples' right to adequate food, right to have safe food, and so forth? How does the current international food law framework address the issue of food waste and its impact on food insecurity? Do the contemporary international food law addresses the obligations of the States and responsibilities along with the liabilities of others in governing the modern food system? Does it address common but differentiate responsibilities of the multi-actors involved in food system? Is the contemporary international food instrument governing a sustainable and equitable global food system, from developing and developed countries perspective? Does the contemporary international food instrument create disparity between developing and developed countries? Is there a need of comprehensive international food law? What are the main components and principles of *the Law of Food Security* deriving from existing international law and norms? How can *the Law of Food Security* contribute to enhancing global food governance by fostering coordination, cooperation, and coherence among various actors and regimes?

The chapter is divided into five segments including introduction and conclusion. The discussion begins with examining the contemporary international food laws under three categories. Then elucidating the need of international food law in building a sustainable global food system, the chapter concludes by asserting the need of international food law.

5.2 Analyzing Contemporary International Food Law Using RIT Model

Enactment of legislation, policy, or regulations conventionally develops a two-party relationship between the rule-maker, or regulator (R), and the rule-taker or target (T) (Abbott et al. 2017). However, when it comes to the formulation of any legislation, policy, or regulations that will have global effectuation forms a three-party relationship, i.e. between regulators (R), intermediaries (I), and target (T), where intermediaries play vital roles from providing expertise feedback to facilitating implementation, and monitoring. Any public or private sector actors, such as NGOs, governmental bodies, trans-governmental organizations, international organizations, or any other organizations of such nature, can be intermediaries.

RIT model is used to analyze international food law for three reasons: firstly, this method is helpful for exploratory studies focusing on the legal mechanism, i.e. set of rules, or procedural steps, addresses "how" questions. Secondly, it helps to analyze the law from the perspective of acceptance, implementation, and purpose achievement. And thirdly, it contributors in developing novel theoretical arguments for problem identification and reform of the existing law. The law or regulation (R) is first analyzed individually, and then its relation with the target (T) is examined. Later, the role of intermediaries in the process and progress of reaching the target is analyzed. The RIT model helps to define the regulation, analysis objective, and take an analytic approach to understanding the outcomes (Fig. 5.1).

The historical analysis of the development of the contemporary international food law and the contemporary global food system suggests three leading categories of international instruments relevant to food security: food safety, standards, and rights. These categories are examined through three major international instruments:—(1) Codex Alimentarius; (2) SPS and TBT Agreement; and (3) UN General Assembly Resolutions. Using the RIT model, these international instruments governing the contemporary global food system are analyzed from the perspective of building food security, general principles, rights and obligations, and their role in global food system. Table 5.1 presents the analysis of these international instruments using RIT model.

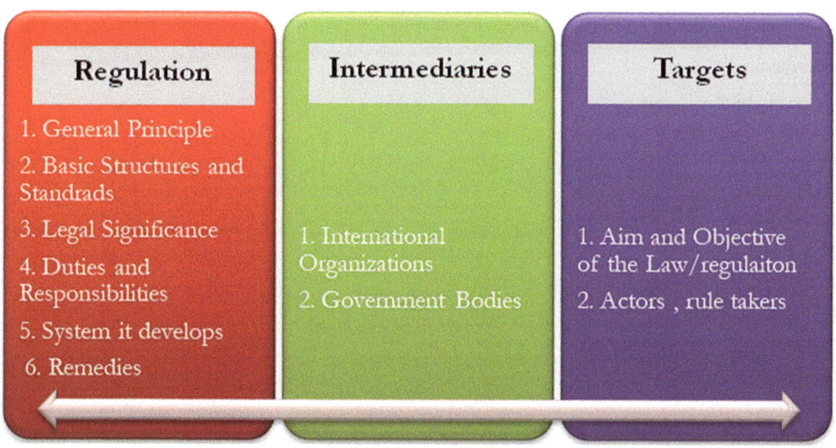

Fig. 5.1 Analyzing law/regulation using RIT model. (Source: Author Developed)

Table 5.1 RIT model analysis of contemporary international food instruments

	Regulations			
Regulators	Codex Alimentarius	SPS Agreement	TBT Agreement	UN General Assembly Resolution
FAO	√			
WTO		√	√	
WHO	√			
UN				√
Substantive rules		√		
Procedural rules	√	√	√	
General and legal principles				
Basic structure				
International standards	√	√	√	
Legal significances		√	√	√
Sustainable consumption of food/anti-food waste				
Carbon neutrality				
Duties and responsibilities	√	√	√	√
Remedies/dispute settlement		√	√	
Risk management				
Intermediaries				
Governmental bodies	√	√	√	√
Other international organization	√	√	√	√
Rule taker and target				
Developed countries	√	√	√	√
Developing countries		√	√	√
Food security				
Food safety	√	√	√	
Food standards	√			√

(continued)

Table 5.1 (continued)

	Regulations			
Regulators	Codex Alimentarius	SPS Agreement	TBT Agreement	UN General Assembly Resolution
Right to food				√
Right to adequate food				√
Right to have safe food	√	√		√
Food related rights				
Ensures enforcement of rights				
Four pillars of food security				√
Food aid	√			√
Global food management system				
Institutional framework (Global food system)				

Source: Author Analysis

The RIT Model analysis suggests a plurality of intermediaries, including governments and international organizations. However, the Codex Alimentarius, SPS and TBT Agreements share a common target to ensure food safety, particularly during international food trade. Moreover, the UN General Assembly Resolutions act as scattered instruments without creating an institutional framework to govern the contemporary global food system to build food security. The RIT model analysis shows that the international instruments governing the contemporary global food system are based on food safety only and are operating without any global food management system and proper institutional framework. Hence, the issue of food security is not being adequately addressed, and food insecurity remains a current global issue that is clandestinely exacerbating. The detail of the analysis is discussed in Table 5.1.

5.2.1 RIT Model Analysis of Codex Alimentarius

Beginning with the findings of RIT model analysis of the Codex Alimentarius, created by the UN's FAO and the WHO, it is apparent that it serves as an important tool for ensuring food security worldwide. The findings suggest that Codex focuses mainly on food safety and standards, providing a framework for countries to set standards for their own food safety and consumer protection systems. The Codex also provides guidelines for international food trade to ensure that all food purchased and imported from countries that adhere to the Codex is safe and of good quality. Additionally, the Codex assists countries in developing their own food safety systems by providing a solid foundation of standards that can be used to design and implement nationally recognized rules and regulations. This helps to ensure that food within a country is of a safe and acceptable standard, and that food imports meet the same standards. As a result, the Codex has developed over 300 Codex standards, guidelines, and other recommendations relating to food quality composition and safety. Such as, *Codex Code of Practice General Principles of Food Hygiene,*[1] *Code of Practice on Food Allergen Management for Food Business Operators,*[2] *Code of Practice for Fish and Fishery Products,*[3] *Code of Hygienic Practice for Low-Moisture Foods,*[4] *Regional Code of Hygienic Practice for Street-Vended Foods in Asia,*[5] *Code of Practice for the Prevention and Reduction of Mycotoxins in Spices,*[6] and so forth. While these international codes are not strictly binding in nature, they are not completely without legal significance, as they are aimed at developing and harmonizing food safety and standards worldwide. The Codex Alimentarius outlines are designed to provide the right balance between public health protection and enabling safe and fair trade of food products. This is achieved by establishing general requirements for food safety, such as good manufacturing practices, hygienic handling practices, labeling requirements, and limits on contaminants and additives. A major accomplishment of the Codex is providing an international standard for assessing nutrition labelling, which can ensure consumers receive accurate information about their food. The Codex also directs food business operates (FBO) to adhere to principles such as good hygiene practices (GHPs) and good manufacturing practice (GMP).

Furthermore, the Codex supports a number of emergency initiatives, such as food aid, export certificates, and recall procedures. This can ensure that emergency scenarios are managed in a manner that protects public health and the environment. They issue guidelines and recommendations on agricultural and fishery products, foodstuffs, food additives and contaminants, feedstuffs, veterinary drugs, pesticides, including labeling, methods of analysis and sampling, codes of ethics and good agricultural practice, and guidelines of hygiene practice, in order to protect consumers' health and ensure fair practices in international trade (Wood 2004).

---

[1] CODEX. Codex Code of Practice General Principles of Food Hygiene [S] 2020.

[2] CODEX. Code of Practice on Food Allergen Management for Food Business Operators [S] 2020.

[3] CODEX. Code of Practice for Fish and Fishery Products [S] 2019.

[4] CODEX. Code of Hygienic Practice for Low-Moisture Foods [S] 2018.

[5] CODEX. Codex Alimentarius : Regional Code of Hygienic Practice for Street-Vended Foods in Asia [S] 2017.

[6] CODEX. Codex Alimentarius : Code of Practice for the Prevention and Reduction of Mycotoxins in Spices [S] 2017.

Although Codex Alimentarius is voluntary in nature and needs to be translated into national legislation or regulations to be enforceable, it has gained importance with the increasing international food trade. This is mainly due to the encouragement of its use as a benchmark against which national food regulations are evaluated, within the legal parameters of the WTO's SPS and TBT Agreements. However, under WTO influence, the Codex has become more "politicized" due to the significant trade interests at stake in the standard-setting processes.[7] Firstly, Codex standards require not only the use of advance technology but also technological innovations, which requires investments and financial assistant. Secondly, Codex standards often exacerbate food waste, which leads to food insecurity. Lastly, although the Codex standards aim at harmonize food safety and standards, they are not developed considering the perspective of developing countries. For instance, *Regional Code of Hygienic Practice for Street-Vended Foods in Asia*, which is a region with many developing, prescribes the use of "equipment, including containers, made of materials that do not transmit toxic substances, odour, or taste, are not absorbent and will not result in food contamination."[8] Manufacturing and purchasing of such equipment and containers are often not economically conformable for many developing countries. Furthermore, although the Code addresses Street-Vended Foods in Asia, it does not take into account the living standard of the people in developing countries, nor does it suggest how developing countries can offer an optimum level of food safety. Similarly, many Codex standards with limited resources can provide an optimal level of food safety. Similarly, many Codex standards do not guide developing countries on how to use limited resources to ensure an optimal level of food safety for their people. Although Codex standards are not drafted with developing countries in mind, FAO and WHO have capacity building programs to help developing countries comply with Codex Alimentarius standards and improve food quality and safety by revising food laws and regulations in accordance with Codex Alimentarius. The finding of the RIT model analysis of Codex Alimentarius suggest that Codex standards are public health preventative measures that focus on food security in terms of consumption of safe imported food.

[7] JOINT FAO/WHO Food Standards Programme, FAO & WHO, CODEX Alimentarius Common, Procedural Manual (21st ed.) [S] 2013; See also, Lin, Ching-fu, Scientification of Politics or Politicization of Science: Reassessing the Limits of International Food Safety Lawmaking. Columbia Science and Technology Law Review 2013, 15(1).

[8] Codex Alimentarius : Regional Code of Hygienic Practice for Street-Vended Foods in Asia [S] 2017.

5.2.2 *RIT Model Analysis of SPS Agreement and TBT Agreement*

The RIT model analysis of the SPS and TBT agreements is presented in four segments. The first segment focuses on the reasons for the adoption of the agreements and how regulations are treated under them. The second segment examines the general principles and legal principles of these agreements. The third segment discusses the general scope of the SPS and TBT agreements. Finally, the fourth segment deals with dispute settlement and institutional matters under these agreements. The discussion begins with an examination of the aim and objective of SPS and TBT agreements.

The SPS agreement is designed to guarantee food safety and access to nutritious food. It provides a platform for countries to establish food safety regulations that correspond to the required level of protection adopted by countries, often with regard to the general health of people in the country.[9] However, SPS agreement does not address food safety issues in a comprehensive and effective manner. Rather, it attempts to provide guidelines for countries to enact food safety laws against each other for the purpose of trade liberalization (Silverglade 2000). On the other hand, the TBT agreement is designed to mitigate technical barriers to trade and ensure that products comply with food safety standards. Countries are allowed to set technical specifications, tolerances, and testing standards depending on the country's food safety standards and preferences.[10] Analyzing the reasons for the adoption of SPS and TBT agreements, a few things are clear. First, both agreements focus on three issues—food safety, facilitating food trade, and harmonization in law. Second, both agreements require advance technology. Consequently, developing countries may encounter special difficulties in "complying with the sanitary or phytosanitary measures of importing Members"[11] and "in the formulation and application of technical regulations and standards and procedures for assessment of conformity with technical regulations and standards".[12] Thirdly, neither agreement addresses food security issues such as food production, food availability and affordability, and market stability. However, contributes in food waste by encouraging measures without complementing food security issues. For instance, Nigeria generates food waste and faces overpricing of food to maintain an "appropriate level of food safety standard" and ensure Hazard Analysis Critical Control Point (HACCP).

Examining the application of general principles within SPS and TBT Agreements, it is observed that the agreements prescribe procedural rules to maintain international food safety standards, which impose duties upon the governments of both developing and developed countries to implement food safety measures. Although SPS and TBT Agreements aim at harmonizing international food safety standards, the

[9] SPS Agreements [S] 1995.

[10] TBT Agreements [S] 1995.

[11] SPS Agreements [S] 1995.

[12] TBT Agreements [S] 1995.

agreements allow governments to regulate a higher standard than an international benchmark provided that it complies with additional criteria such as a "risk assessment" (Kurtz 2007). This results in contrasting international food safety standards, which vary between developing and developed countries. It is observed that food safety standards of developed countries are often more stringent than those of developing countries. Due to contrasting food safety standards, developing countries often lose opportunities to access the market. This indicates that food safety standards are conflicting and are the result of conflicting principles, such as principle of harmonization of law, principle of equity, and principle of standardization, in the SPS and TBT Agreements. Although special and differential treatment (SDT) is among the major principles that SPS and TBT Agreements acknowledge, it refers to offering a longer transition period to the developing countries with some limited flexibility to cope with the economic pressures of trade liberalization (Keck and Low 2011). Moreover, the SDT principle offers a set of legal provisions and rights (Hegde and Wouters 2021) such as, supporting developing countries to build infrastructure (Ayoki 2018) handle disputes (South Centre 2017), and implement technical standard that require documentation to establish compliance with the SDT principle. However, the SPS Agreements do not require Members to document how they have complied with SDT principle,[13] which upsets the balance of rights and obligations between developed and developing country.[14] Furthermore, it is also observed that as the SDT principle is not acknowledged in a broader aspect, developed countries often do not take the needs of developing countries into account (Ayoki 2018; Surono and Hidayati 2019). Further analyzing the general principles acknowledged by both agreements, it is observed that the agreements acknowledge the principle of non-discrimination, the obligation to prevent unnecessary obstacles to international trade, and the obligation to use international standards as a basis for technical regulations (Disdier et al. 2008). However, the exception clauses have also provided the liberty to countries to impose higher level of protection measures according to their preference. This sometimes leads to discrimination and creates economic pressure on developing countries.

Considering the general scope of SPS and TBT Agreements, they have limited themselves to food safety and technical standards during international food trade; however, they encourage the development of "National Food Safety System." This applies to both importing and exporting countries, which can be either among developing and developed countries. Additionally, it is observed that the agreements are negotiable in nature (Disdier et al. 2008) as measures under the agreements are often applied based on bilateral agreements. Therefore, it can be asserted that the two agreements address the impact of standards as a mechanism of the public regulatory process on trade flows, thus encouraging bilateral negotiable agreements among the parties to determine food safety standard. Here, it is important to understand whether the agreements encourage measures that are fair and reasonable within the contemplation of both developing and developed countries. As the TBT Agreement

[13] Panel Report, DS293 European Union (formerly EC)—Measures Affecting [S] 2003.

[14] Panel Report, DS447 United States—Measures Affecting the Importation of Animals, Meat and Other Animal Products from Argentina [S] 2015.

focuses on technical regulations and standards to maintain a certain level of transparency regarding food safety, more emphasis is given to disputes regarding the SPS Agreement in the following section, as it particularly focuses on food safety measures.

To comprehend the general scope of SPS Agreements, as well as their acceptance and implementation, it is necessary to analyze WTO cases. To-date, 49 cases have been brought under different provisions of the SPS agreement, of which 23 cases are still ongoing and 26 cases have been decided. Analyzing the 26 decided cases, it is observed that although the SPS agreement aims to harmonize international food safety standards, it encourages private food safety standards by defining the term "appropriate level of sanitary protection" as "the level of protection deemed appropriate by the Member establishing a sanitary measure to protect human life or health within its territory."[15] This indicates that there is no adequate, harmonized or benchmark for food safety standard. However, to achieve harmonization, the SPS prescribes the development of food safety standards "based on international standards, guidelines or recommendations, where they exist."[16] This indicates that the SPS agreement "encourages the harmonization of SPS measures on the basis of international standards, while at the same time recognizes the right to determine appropriate level of protection."[17] Additionally, it is observed that the SPS Agreement acknowledges international standards for food safety, including the standards, guidelines, and recommendations established by the Codex Alimentarius Commission relating to food additives, contaminants hygienic practice. Thus, it can be asserted that the general scope of SPS Agreement is particularly confined to food safety standards. Although SPS Agreement does not consider "(i) whether the standards reflect levels of protection or sanitary measures or the type of sanitary measure they recommend, or (ii) whether these standards have been adopted by consensus or by a wide or narrow majority, or (iii) whether the period during which they have been discussed or the date of their adoption was before or after the entry into force of the SPS Agreement."[18] However, by acknowledging the standards of Codex Alimentarius, these standards have "made the *Codex Standards commercial*[19] and *political in nature* (Johnecheck 2008). The analysis of SPS and TBT Agreements suggests that the objective is similar to the Code Standards in ensuring public health security without harmonization between science, law, and socio-economic aspects. The RIT model suggests that the SPS and TBT agreements provide public health preventative measures that focus on food security in terms of the consumption of imported food.

[15] SPS Agreements [S] 1995.

[16] Ibid.

[17] Appellate Body Reports, DS320/DS321 United States/Canada—Continued Suspension of Obligations in the EC—Hormones Dispute [S] 2008.

[18] Panel Reports, DS321 Canada—Continued Suspension of Obligations in the EC—Hormones Dispute [S] 2008; See also, Panel Reports, DS320 United States—Continued Suspension of Obligations in the EC—Hormones Dispute [S] 2008.

[19] Frode Veggeland & Svein Ole Borgen. Changing the Codex: The Role of International Institutions [R] Norwegian Agric. Econ. Res. Inst, 2002.

5.2.3 *RIT Model Analysis of UN General Assembly Resolutions and Other Instruments*

Analysis of UN General Assembly Resolutions under RIT model suggests that they are a formal expression of opinion, intention, or decision by an official body or assembly holding a normative value.[20] Therefore, the stance of UN General Assembly Resolutions on developing food security is essential. Several major resolutions addressing food-related issues have been analyzed, such as, A/RES/70/259 United Nations Decade of Action on Nutrition (2016–2025). This resolution calls upon the Food and Agriculture Organization and the World Health Organization, in collaboration with other organizations, to lead the implementation of the United Nations Decade of Action on Nutrition (2016–2025) and to develop a work program based on the Rome Declaration and its Framework for Action. Similarly, UN General Assembly Resolutions A/RES/70/154 The Right to Food is another international instrument with legal significance. It acknowledges the right to food as a human right, and UN member states have promised to take adequate measures to protect this right. The UNGA resolution addresses the need for increased food production in order to feed a growing global population and promote better nutrition. It also seeks to provide policy support to improve the resilience of agricultural systems in the face of climate change and other environmental challenges. The resolution specifically targets countries that are most at risk of food insecurity, including low-income nations in sub-Saharan Africa, as well as small island states in the Caribbean. This level of support should enable the production of more nutritious food and increasing the amount of food available to those in need. Furthermore, UN Special Rapporteur on the Right to Food appointed under resolution 2000/10 has contributed largely in the growing concern regarding right to food by defining it as "the right to food is the right to have regular, permanent and unobstructed access, either directly or by means of financial purchases, to quantitatively and qualitatively adequate and sufficient food corresponding to the cultural traditions of the people to which the consumer belongs, and which ensures a physical and mental, individual and collective, fulfilling and dignified life free from anxiety."[21] Furthermore, UN Special Rapporteur addressed the inequality, and the inequitable distribution of food and productive resources, and right to food in international trade. However, these reports of UN Special Rapporteur lack legal significant but it forms a strong ground for the government to ensure "right to food."[22] The resolution also focuses on the need to improve access to food and nutrition in developing countries, placing emphasis on expanding access to markets, credit, inputs, and information to support local farmers and small-scale producers. Additionally, it aims to enhance the global distribution of food, with a particular emphasis on reducing losses and waste throughout the supply chain. Finally, the

---

[20] Alsharqawi, Ahmad et al. The Role of General Assembly Resolutions to the Development of International Law [J] Journal of Legal, Ethical and Regulatory Issues, 2021, 24(2).

[21] UNESC. Report by the Special Rapporteur on the right to food UNESC, E/CN.4/2001/53 [S] 2001.

[22] See, UNGA. Right to food A/74/164. GA Resolution 73/171 [S] 2019.

resolution includes provisions to ensure food safety, such as standards to protect nation's ecosystems and soil quality.

The resolution's direct approach to improving global food production, availability, and safety is a significant step forward for global food security. The directive will have a positive impact on food production, distribution, and consumption, providing a global framework that seeks to support those most at risk of food insecurity while supporting sustainable agriculture and food systems.

Furthermore, the Universal Declaration on the Eradication of Hunger and Malnutrition (UDEHM) is another international instrument with legal significance securing "the right to be free from hunger" as an inalienable right. "It states, "every man, woman and child has the inalienable right to be free from hunger and malnutrition in order to develop fully and maintain their physical and mental faculties."[23] Under this document the Governments are obliged to work together for higher food production and a more equitable and efficient distribution of food between countries and within countries; making it a fundamental responsibility of Governments. According UDEHM, all developed countries and others should collaborate technically and financially with the developing countries. Under UDEHM, ensuring the availability of food at all times at fair cost and cooperation among the countries to establish an effective system of world food security is a common responsibility of the entire international community.

Moreover, UN General Assembly Resolutions have bestowed responsibilities and duties upon the global community to ensure right to food and food security. Although UN General Assembly Resolutions hold certain legal significance, as theoretical analysis of General Assembly resolutions suggests that they are source of international law (Bleicher 1969); mainly due to their notable legal impact in the field of customary law and developing international law.

The RIT Model analysis suggests that UNGA resolutions are a positive step toward improving global food security, providing a clear framework for improving food production and access, as well as supporting more sustainable food systems. Although the directive is only a first step, it serves as a reminder that collective action is needed to ensure that all have access to nutritious and safe food.

However, according to the RIT model analysis in Table 5.1, contemporary international instruments related to food law do not focus on food security, but rather on food safety standards, and lack legal significance and general principles. The analysis confirms that there is currently no "international food law," as the objective of these international food law revolves around food consumption and the notion food safety. The contemporary global food trade system is mainly governed by SPS and TBT agreements, focusing particularly on trade and food safety standards in international trade. Therefore, there is no adequate food management system that works for food management and control and can be addressed as global food system. Additionally, there is lack of uniformity in food safety standards, resulting in disparity between the developing and developed countries. Furthermore, the contemporary

[23] UNGA, World Food Conference. Universal Declaration on the Eradication of Hunger and Malnutrition. UNGA Resolution 3180 (XXVIII) [S] 1974.

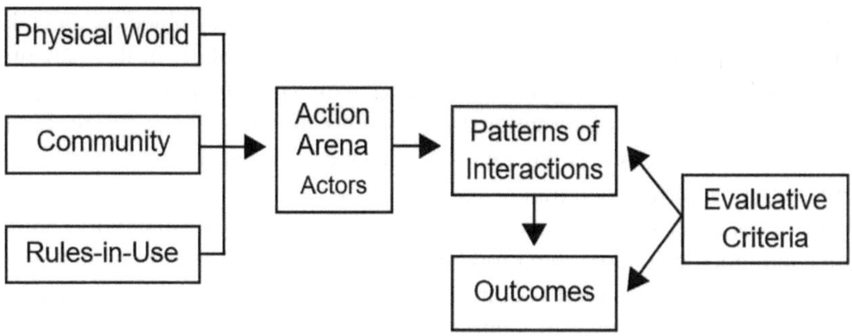

Fig. 5.2 Process of institutional analysis using the IAD framework. (Source: Adapted from Ostrom et al. (1994, 37))

international instruments fall short in building an institutional framework, as they do not ensure the enforcement of food rights or impose responsibilities and obligations. More in-depth analyses of institutional framework and governance are discussed in the next section.

5.3 Analyzing Contemporary International Food Law Using IAD Framework

IAD framework includes analyzing actors, norms, institutional settings, incentive structures, rules, and more. Social researchers have widely used the model for policy analysis (McGinnis 2011) especially for answering—such as what is likely to happen if current policy A would be replaced by policy B, and what needs to be done to implement B? The IAD framework is useful for institutional analysis and policy design, and is appropriate for this study because it analyzes the contemporary global food system and aims to develop a legal mechanism and institutional framework to reform it into a sustainable and equitable global food system. There are three reasons for using IAD framework: firstly, it helps to evaluate, design, or reform policy by systematically analyzing existing arrangements and t generating and comparing alternatives."[24] Secondly, it offers more comprehensive and precise policy analysis, which helps to carefully consider a wide range of issues that are important aspects of a particular policy problem, thus offering successful policy solutions. Lastly, as the framework demands multiple disciplinary perspectives, it holds the potential to provide a basis for more effective policy by identifying the consensus for coordinated action. Figure 5.2 shows the process of institutional analysis using the IAD framework.

[24] Polski MM, Ostrom E. An institutional framework for policy analysis and design [M] In Cole DH, MD MG (eds.) Elinor Ostrom and the Bloomington School of Political Economy: A Framework for Policy Analysis. Lanham: Lexington Books, 1999: pp 13–47.

The first step of IAD framework involves defining the policy objective and taking an analytical approach to understand the outcome of the policy. In this part of the analysis the findings of the RIT model are used, which clearly defines contemporary international food law as international instruments that are neither strictly binding in nature nor completely lacking legal significance. These instruments govern food safety standards during food trade (see Table 5.2).

In the second step of IAD framework, we analyze physical and material conditions of the contemporary international food law. Following standard economic theory, it is observed that the economic nature of these instruments is high subtractability and excludability. The level of subtractability and excludability of these international food law is analyzed based on the standard of food safety rules, the need for technology to maintain them, the scale and scope of activity required, and other aspects such as active participation. The high subtractability and excludability indicates that the contemporary international food law offers food safety standards that are not comfortably affordable by developing countries. Furthermore, it causes enormous quantity of food waste and exacerbates food insecurity. In the third step of IAD framework, we analyze community attributes, rule-in-used, action arenas, and patterns of interaction that suggest the policy action situation, acceptance of norms, policy activities, and the extent to which potential participants' values, beliefs, and preferences about policy-oriented strategies. For the purpose of the analysis using IAD framework, two categories of international instruments are considered: Codex Alimentarius, and SPS Agreement (Table 5.3). It is observed that the contemporary international food law offers certain food safety standards, and to maintain such safety standards, advanced technology, technological innovation, research and infrastructure, including the establishment of national regulatory bodies, technical expertise, training, and equipment, among others are essential. This requires investment, financial support, grants, and donations. This also indicates that the contemporary international food law is not offering equal opportunities for developing countries to exports (Wilson et al. 2001).

The preamble of SPS Agreement acknowledges that "developing country Members may encounter special difficulties in complying with the sanitary or phytosanitary measures of importing Members."[25] This indicates that food safety standards are not drafted with the conditions of developing countries in mind, and developing countries are not given guidance on how to use their limited resources. Instead, it offers developed countries the privilege to exert economic pressure on developing countries in certain cases (Alam and Tomossy 2017; Das 2008). Under the SPS Agreement, "members agreed to facilitate the provision of technical assistance to other Members, especially developing country Members, either bilaterally or through the appropriate international organizations."[26] However, this provision also allows developing countries to become dependent on developed countries. Moreover, the SPS Agreement also requires developed countries to offer longer preparation and application periods for sanitary or phytosanitary measures, taking into account the

[25] SPS Agreements [S] 1995.

[26] SPS Agreement [S] 1995.

Table 5.2 Policy definition and objective

	Regulations			
Policy definition	Codex Alimentarius	SPS Agreement	TBT Agreement	UN General Assembly Resolution
Substantive rules		√		
Procedural rules	√	√	√	
General and legal principles				
Basic structure				
International standards	√	√	√	
Legal significances		√	√	√
Sustainable consumption of food/anti-food waste				
Carbon neutrality				
Duties and responsibilities	√	√	√	√
Remedies/dispute settlement		√	√	
Risk management				
Objectives				
Food security				√
Food safety	√	√	√	
Food standards	√			√
Right to food				√
Right to adequate food				√
Right to have safe food	√	√		√
Food related rights				
Ensures enforcement of rights				
Four pillars of food security				√
Food aid				√

(continued)

Table 5.2 (continued)

	Regulations			
Policy definition	Codex Alimentarius	SPS Agreement	TBT Agreement	UN General Assembly Resolution
Global food management system				
Institutional framework (Global food system)				

Source: Author Analysis

financial, trade, and development needs of developing countries. However, this also suggests that maintaining an appropriate level of sanitary protection will cost developing countries more, which will ultimately increase the cost of food price and affect its affordability. Additionally, it may also impact the food production line of the developing countries. Hence, the participation of developing countries is not satisfactory, and the disparity between developing and developed countries is visible. This further substantiates that contemporary international food law focusing particularly on food safety during international food trade falls short of building a global food system and lack an institutional framework. Currently, some scattered the international food law are working, offering a legal mechanism without forming any institutional framework and governance. Hence, it is essential to develop international food law to address these issues.

5.4 Global Food Governance Model

With the increasing population of the world and a growth in the demand for food, it is essential for governments and stakeholders to collaborate and develop a global food governance model. The current food systems have proved to be inadequate in terms of providing and regulating enough food to meet the needs of the population, while preserving the environment. A successful global food governance model should consider multiple aspects of the food industry, particularly in terms of production, distribution, and access.

Global food governance model is a framework to developing a set of rules and regulations to govern the production, storage, trade, and distribution of food-related products and services on a global scale. It also influences policies related to food security, nutrition, and public health. Food governance plays an integral role in ensuring that the world's food supply is adequately managed, distributed equitably, and remains affordable for everyone. This can be achieved through the establishment

Table 5.3 IAD analysis of Codex Alimentarius and SPS Agreement from the perspective of Food Security

IAD domain	Codex Alimentarius	SPS Agreement
Physical world	Provides multiple documents suggesting food safety standards	Provides the right to set appropriate level of food safety standards
	Induces food waste	Induces food waste
	Food insecurity exists	Food insecurity exists
		Contributes in food availability
Community attribute	Developed countries satisfactory level of participation	Developed countries are more engaged compared to developing countries
	Developing countries poor level of participation	Often appropriate level of protection is set as standard to create economic pressure
		Many developing countries struggle to apply SPS measures
	Heterogeneous individualistic	Heterogeneous individualistic
Rules-in-use	Lack of legal significance to adhere the standards	Privileged to determine appropriate level of food safety standards
	Determines food safety standards	Determines food safety standards during food trade
		Facilitates food trade
Action arena	Developed countries	International food trade
	Developing countries struggles	Makes food available by food trade
	International Organizations actors as intermediaries	
Patterns of interaction	Developed countries can adhere to	Disparity between developing and developed countries.
	Developing countries struggles	Often used to create economic pressure
Outcomes	Ununified and discrepant food safety standards forming disparity between developing and developed countries and does not address food security and food management issues	

Source: Author Analysis

of mechanisms such as "Food Price Monitoring and Analysis" or "Retail Pricing Laws and Regulations."

A global food governance model should address a range of issues, including food security, nutrition, agricultural trade and investment, and the social and economic impacts of food production. The goal of such a model is to develop a "Sustainable and Equitable Global Food System" (see Fig. 5.3) that ensures the availability of

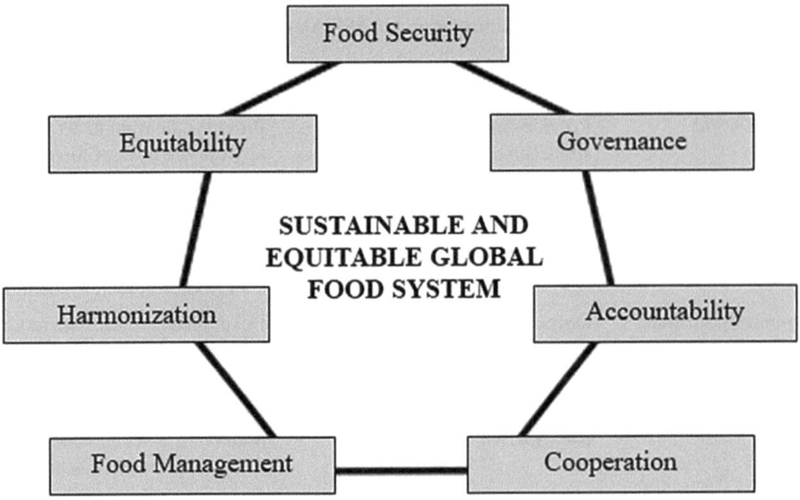

Fig. 5.3 Benefits of sustainable and equitable global food system. (Source: Author Developed)

food, access to safe and nutritious food, affordability of food, equitable distribution of food, sustainable consumption of food, and stresses on food management.

It is important to clarify the difference between food governance and food management in the context of Fig. 5.3. While the two terms are often used interchangeably, they represent distinct concepts. Food governance is concerned with policymaking at all levels, including the establishment of production standards and safety requirements for the agricultural industry through policy, legislation, and regulations. Food management, on the other hand, focuses on the operational aspects of large-scale food systems, such as procurement processes, inventory control systems, quality assurance programs, production planning and scheduling functions, transportation services optimization efforts, and other activities. Additionally, the Sustainable and Equitable Global Food System presented in Fig. 5.3 emphasizes both harmonization and cooperation as methods of working together. Harmonization involves bringing multiple entities into agreement, recognizing that each may need to modify their policies, procedures, and structure to achieve a common goal. Cooperation, on the other hand, focuses on leveraging the strengths and resources of all entities involved to advance a shared objective.

The first essential factor in any global food governance model is the promotion of food security. This involves ensuring the availability of adequate safe and nutritious food, as well as the ability to access it sustainably. This can be accomplished through measures such as improved agricultural infrastructure, agricultural trade reforms, and improved policy for food safety.

The second essential factor is the promotion of food safety. This is related to food insecurity, but focuses more on the safety of food production, storage, and preparation. This can be addressed through better food safety regulations and monitoring systems.

The third essential factor is responsible investment in agriculture. This involves encouraging the sustainable use of natural resources, such as biodiversity, water, and land, and promoting the growth of small-scale agricultural production. Investment in agricultural technology can also improve food production efficiency and reduce the environmental impact of food production.

The fourth essential factor is the promotion of nutrition education. This includes raising awareness about the importance of consuming nutritious food and providing education and training on proper nutrition to improve overall public health. Nutrition education should also include guidance on how to best use the available food resources and information on the nutritional benefits of different foods.

The fifth essential factor is the promotion of equitable access to food. This involves addressing social and economic inequalities that may hinder access to food for certain populations, such as low-income individuals or marginalized communities. Measures to promote equitable access to food may include food assistance programs, fair trade policies, and addressing issues related to food waste and distribution.

The sixth essential factor is the promotion of sustainable food production and consumption. This involves minimizing the negative impact of food production on the environment, promoting sustainable farming practices, and encouraging sustainable food consumption habits, such as reducing food waste and choosing plant-based diets.

The seventh essential factor is the establishment of effective governance structures and institutions. This includes creating regulatory frameworks, establishing monitoring and evaluation systems, and promoting transparency and accountability in the global food system. Strong governance structures and institutions are essential for ensuring the effective implementation of policies and regulations related to food security, safety, and sustainability.

Finally, the eighth essential factor is an appropriate legal and policy framework. This includes international or regional agreements and laws that can ensure equitable access to food, protecting the right to food, and prevent the exploitation of vulnerable populations. However, the food governance framework (Alemanno 2015) pertains to a set of challenges, including:

1. The demarcation and coordination between assessment and management of food safety threats;
2. The handling of scientific uncertainty;
3. The increase of transparency during the entire food safety governance process;
4. The involvement of a diversity of social groups and the wider public into the governance process;
5. The handling of highly controversial food safety issues.(Renn and Dreyer 2009)

Hence, the first step in developing effective global food governance is to recognize the importance of local and global processes, both in terms of production and food access. On the production side, global food governance should involve coordinating investments in research, agronomy, and technology to ensure the production of safe and nutritious food. Additionally, incentives should be developed to encourage investment in sustainable practices, such as organic agriculture and conservation of natural resources. On the access side, global food governance should consider strategies for increasing food access to vulnerable populations and minimizing food waste, particularly when access to food is at stake. According to the UNHCR, "around 100 million people are displaced globally and must rely on cash and food assistance to survive."[27] Often, these people find themselves in precarious positions, with limited access to food and other basic needs. Refugee camps, for example, which may lack infrastructure and food distribution networks, can leave those living in them vulnerable. In some cases, with little to no control over their income or work opportunities, these populations may face long-term food insecurity and malnutrition.

A second example of non-accessibility of food is in the case of natural disasters. When a natural disaster strikes—such as a tsunami, monsoon, hurricane, or drought—it can cause severe food insecurity among populations. In the wake of any natural disaster, access to food is severely limited or not available at all. Those affected by the disaster can face challenges in locating and transporting food due to the destruction of roads, ports, and other infrastructure ("Effects of Natural Disasters"). If food cannot be accessed, these populations have few options and are susceptible to even greater food insecurity.

A third example of non-access to food could be when people live in a food desert. A food desert is an area in which people have "limited access to affordable and nutritious fresh food, often due to poverty or geographic isolation" ("Food Deserts"). This is particularly common in inner cities or rural areas that can lack grocery stores and restaurants that offer healthy, affordable food options. As a result, people in these areas are left with few good food choices, and as such, can suffer from malnutrition or obesity.

Food governance can improve non-access of food situation through several ways. Firstly, policy frameworks that incentivize both production and access, helping to reduce poverty and recognize the social and economic value of food. Second, leveraging existing public infrastructures, such as public transport, can ensure that even remote and rural communities gain access to adequate nutrition. Lastly, empowering local communities and engaging in mutual learning and dialogue can also help improve access to food.

To ensure accountability and compliance, the development of a global food governance model should include mechanisms for monitoring and evaluating implementation. These mechanisms should consider potential risks, such as climate change and its effect on food security, as well as the potential for exploitation of vulnerable populations. Additionally, international treaties should be negotiated and signed to ensure adherence to agreed-upon standards. Compliance should be assured through a robust regulatory framework, involving multiple governments, international organizations, and stakeholders.

To achieve the desired outcome of access to safe and nutritious food, a global food governance model should also involve consumer participation in decision-making. This should include initiatives to promote consumer awareness regarding

[27] UNHCR: A Record 100 Million People Forcibly Displaced Worldwide [N] UN News 2022-05-11-03.

food production and its effects on their health and the environment. It should also include mechanisms to increase consumer access to information and education, and ultimately, involve consultation with consumer groups prior to any policy decisions.

Seeing that global food governance is a complex undertaking, adequate resources must be allocated to ensure effective implementation. This involves collaboration between countries and other stakeholders to access the funds necessary for research, development, and implementation of effective global food governance measures. It is also essential for representatives from all affected populations to be actively involved in the decision-making process to ensure that the resulting model is equitable and serves the interests of all relevant parties. A successful global food governance model should take into consideration the local and global processes involved in the production, distribution, and access to safe and nutritious food. Furthermore, a robust legal framework, monitoring and evaluation mechanisms, and adequate resources must be established. Additionally, consumer involvement in decision-making should be encouraged to ensure the necessary public engagement and accountability for the project. In the below section basic structure of *the Law of Food Security* is discussed.

5.5 Basic Structure of the Law of Food Security

The idea of law is the utmost principle of law-making and law-operating, where, the idea of law is the rational cognition to the essence and development rule of law, which forms the fundamentals to establish legal institution and legal regulation mechanism. Zhang Lili stated the concretization of the idea of law and the concentrated expression of the rules of legal system constitutes the basic structure of law (Zhang Lili 2011). So, it can be asserted that the basic structure of any law consists of the foundational structure of the law such as, the aim and objective of the law which develops a legal system that constitutes rights, privileges, obligations, and accountability. Hence, before outlining probable and essential provisions, it is essential to draft the basic structure for the Law of Food Security, which will guide in identifying and developing essential provisions to be included in the Treaty.

Food issue is an extremely important issue and the food security issue is an extremely important security. So, *the Law of Food Security* is another form of international security, which is not a set of rules regulating the maintenance and restoration of international peace and security. Rather contributed largely in maintaining international peace and social order by ensuring the right to food for all at all times. Generally thinking about "food law" it is believed food law will be a legislation which will regulate the food production, food supply chain, food standard, food safety, food trade, and other relevant aspects with an objective to build food security. So, it can be asserted that *the Law of Food Security* must contain an aim and scope with some general principles, rules for food production, safety, trade and labeling, rules to prevent food waste, and rule for food assistance and aid. It means, *the Law*

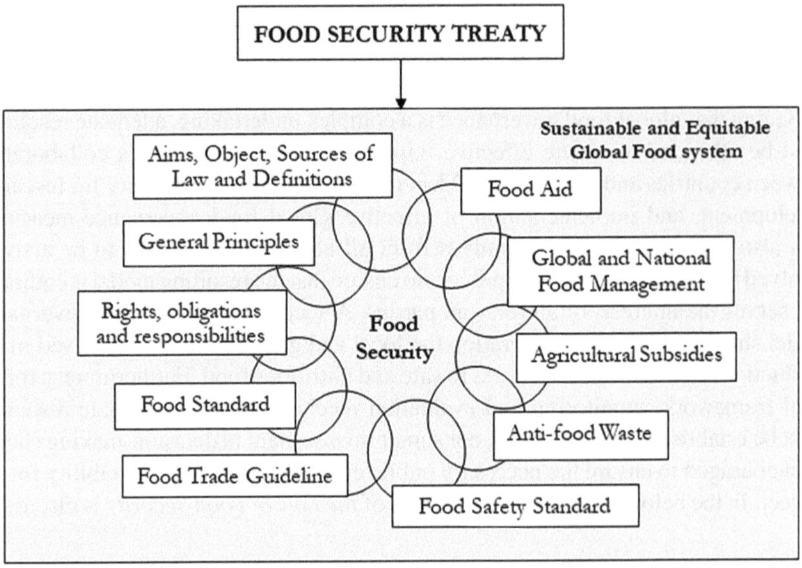

Fig. 5.4 Basic structure for the *Law of Food Security*. (Source: Author Developed)

of Food Security must contain some rights and privileges, obligations, and accountability. In pursuance of enacting *the Law of Food Security* below aspects can be considered (Fig. 5.4).

Beginning with the objective and goal of the Law of Food Security. The objective of *the Law of Food Security* must be preventing the violation of people's right to food because of food insecurity and set its goal at achieving food security to ensure people's right to food. Addressing notions like food insecurity, food security, and right to food, an adequate legal definition of these notions including other relevant notions are also essential. The contemporary international law can act as a source and guidance in this regard. Furthermore, since the concept of *the Law of Food Security* covers a wide range of areas such as, nutrition, food safety, food trade, food standard, human rights, environment, and so forth, identifying the source or basis of international food law is prerequisite to address the broader aspect to international food law and will also make the contemporary international instruments a source of the Law of Food Security. Analyzing the benefit bestowed by Article 38 of the International Court of Justice Statute to International Court of Justice will help to comprehend the benefit of identifying the sources of international food law. Moreover, because of the broader aspect *the Law of Food Security* needs to acknowledge general principles and customary law besides treaties, agreements, and conventions. General principles act as a "gap fillers" and "bridge" between customary law, contemporary law and measures, and creates delimits actions. Implication of principles such as precautionary principle, common but differentiated principle, principle of non-discrimination, equitable and sustainable utilization, due diligence in *the Law of Food Security* also needs further research.

Withal, as food security has four aspects, food production and supply, food standard, food safety, food trade, and food waste. Hence, it will be convenient to divide *the Law of Food Security* into segments. For instance, food safety is considered more as food trade issue rather than food security issue. However, food safety and food standard is more related to food insecurity issue. Because, food security exists when the supplied food is safe and provides adequate number of calories and nutrients.

Implication of food safety at national level barely grasped international concern. Similarly, food production and supply are also considered from international trade aspects and at global levels, but its nexus to maintain food security at national levels is not considered. Nor, any legal framework is suggested for food production and to maintain equitable distribution of food at global and national levels. Hence, many prominent issues such as, food waste during food production, technological and financial support for the developing and least countries, saving farmers and subsides never gained concern at international level. Despite that fact that, without solving these issues at national levels, food security at global level cannot be resolved as with time food waste is gradually increasing with the severity of food insecurity. Hence, it is prerequisite for the international food law to address food waste issue (detail on how food waste exacerbates food insecurity is discussed in Chap. 3).

Moreover, dividing *the Law of Food Security* into segments also suggests the urge of rights, obligations, and responsibilities. Hence ensuring rights such as right to food, right to adequate food, right to have safe food, right to have food of own preferences, and right to have food with adequate nutrients, becomes prerequisite for *the Law of Food Security* along with their adequate illustrations or definitions. However, unlike right to food, right to adequate food, right to have safe food, other relevant rights such as right to have food of own preferences and right to have food with adequate nutrients are not explored in the international arena. But by reading the definition of right to food and food security together we can get support for other relevant rights such as right to have food of own preferences and right to have food with adequate nutrients. However, these rights are on more general in nature, or can be referred as more consumer oriented. But acknowledgment of other rights such farmer's rights for technological support, right to farming, and adequate price for products is also essential.

Accordingly, to ensure any rights acknowledgment of obligations and responsibilities is also essential. Implementation of obligations and responsibilities upon different stakeholders will contribute largely in the enforcement of the Law of Food Security. These obligations and responsibilities will not only help to build a national food system but will also incorporate food governance within the global food system. Hence, for the effective implementation and enforcement of *the Law of Food Security* a "rule review mechanism" is required which will be a peer review process to assist the States parties to effectively implement the Treaty. Establishment of rule review mechanism is not a new concept for the effective implementation of Treaty. A review mechanism is established for the implementation of United Nations Convention against Transnational Organized Crime (UNTOC) and its Protocols by adopting resolution 9/1 entitled *Establishment of the Mechanism for the Review of the Implementation of the United Nations Convention against Transnational Organized Crime and the Protocols thereto.* Later the State Parties also adopted resolution 10/1 entitled *Launch of the review process of the Mechanism for the Review of the Implementation of the United Nations Convention against Transnational Organized Crime and the Protocols thereto.* Furthermore, implementation of obligations and responsibilities inserts liability and accountability which often raises disputes, that needs to be settled down legally in cooperative manner. Hence, *the Law of Food Security* also requires a dispute settlement mechanism, which can either follow the conventional international dispute settlement mechanism such as, WTO dispute settlement mechanism for particular trade relevant issue, or International Court of Justice (ICJ) for contentious cases regarding food security. Either it can hold its own dispute settlement mechanism like the International Tribunal for the Law of the Sea (ITLOS). Additionally, *the Law of Food Security* needs to set up legal framework for risk assessment and risk management, knowledge management, along with food aid and assistance.

The detail of the outline of the provisions discussed in the basic structure is discussed in the details in Chap. 6. As a branch of International Law, *the Law of Food Security* will represent "the will and consent" of nation-states with respect to the rules that will govern the global food system for a sustainable purpose. So, *the Law of Food Security* with its comprehensive and interdisciplinary approach will initiate some common practices, rights and obligations, cooperation, and codes of conduct by creating a system through global food governance which will develop high accountability standards, transparency, documentation of everything, promote and facilitate the exchange of information, initiate good practice, and promote constructive and technical collaboration.

5.6 Negotiating the Law of Food Security: Power Dynamics and Contentious Issues

The issue of food security is a pressing concern that affects millions of people worldwide who are facing hunger and malnutrition. To address this challenge, the proposed Law of Food Security aims to establish binding commitments for countries to work together and combat global food insecurity through coordinated efforts. However, the negotiation surrounding the enforcement of the Treaty has the potential to become a contentious issue, given the ongoing negotiations related to addressing food security through trade policies (MacMaolain 2008). Developed and developing countries may disagree over the power dynamics and contentious issues involved in the negotiation process, which could affect the success of the treaty's implementation.

In this section, the power base for global agreements and the arguments of developing and developed countries in the negotiation of the proposed *the Law of Food Security* is examined. We analyze the potential contentious issues that have historically been a source of disagreement among countries and stakeholders, including trade policies, agricultural subsidies, technology transfer, climate change, and resource allocation. Furthermore, we explore the different perspectives of developing and developed countries regarding these issues and the potential implications of their positions on the negotiation of the Law of Food Security.

5.6.1 The Power Base for Global Agreements

The negotiation of global agreements involves complex power dynamics that shape the outcome of the negotiations. Literature suggest that international organizations, UN and governments play an important role in global agreements. Loibl found that international organizations are no longer just advisors and observers, but are active participants in international law-making, providing expertise in areas such as environmental law (Loibl 2001). Gabriela emphasizes the importance of international organizations in the new paradigm of global economic governance, while Kent discusses the role of international organizations in implementing the principle of policy integration, particularly in trade and investment organizations. Chazournes highlights the evolving role of international organizations and the need for innovative legal mechanisms to adapt to contemporary requirements (de Chazournes 2009). It can be asserted that international organizations are crucial in shaping and implementing global agreements. Similarly, UN also plays an important role in global agreements related to peace, security, and stability. Kanu and Oseni argue that collective agreements are necessary to tackle contemporary threats to world peace and stability, and that a restructured UN Security Council is necessary for a more responsive and all-inclusive organization (Kanu and Oseni 2022). Mingst, Karns, & Lyon emphasizes the UN's role in working on dangerous conflicts among and within states, and provides a brief overview of the UN's mission to maintain international peace and security and to cooperate in solving global problems. The ability of countries to influence and shape these agreements is based on a range of factors, including economic, political, and cultural power (Mingst et al. 2022). Understanding the power base for global agreements is crucial to understanding the dynamics of negotiations and the potential outcomes of these agreements.

The power base for global agreements is complex and multifaceted, involving a range of economic, political, and cultural factors. Understanding the dynamics of power in the negotiation process is crucial to achieving successful outcomes that benefit all parties involved. The negotiation of *the Law of Food Security* is no exception, and requires careful consideration of the power dynamics at play to ensure that the agreement is fair and effective in addressing global food insecurity.

5.6.2 Potential Contentious Issues in the Negotiation of the Law of Food Security and the Arguments of Developing and Developed Countries

The negotiation surrounding the enforcement of the Treaty has the potential to become a contentious issue. Various contentious issues might arise that hold the potential to cause disagreements among developing and developed countries. These issues involve trade policies, agricultural subsidies, technology transfer, climate change, and resource allocation.

One of the major issues in the negotiation process is trade policies. Developing countries have argued that developed countries' trade policies have negatively impacted their agricultural sectors and impeded their progress toward food security. Developed countries, on the other hand, have argued that trade policies should not be restricted, as they are critical for promoting economic growth and reducing poverty. Therefore, there is a need to establish guidelines for trade policies that support the interests of developing countries while also promoting economic growth. Similarly, agricultural subsidies are another contentious issue in the negotiation of the Law of Food Security. Developing countries have argued that developed countries' agricultural subsidies distort the global market, harm their agricultural sectors, and limit their ability to compete in the global market. Developed countries have argued that agricultural subsidies are essential for ensuring food security and supporting farmers in their countries. Therefore, there is a need to establish provisions that restrict agricultural subsidies that distort the global market while also providing support for farmers (Bronfenbrenner 1956; FAO 2021; Micheni et al. 2022; Rivera 2005).

Climate change is also a significant factor in the negotiation of the Law of Food Security. Developing countries have argued that they are disproportionately affected by climate change and need more support to mitigate its impacts on their agricultural sectors. Developed countries have argued that they are already taking significant steps to address climate change and should not be required to provide more support. Therefore, there is a need to establish provisions that promote sustainable agricultural practices that mitigate climate change and enhance agricultural productivity (Gerber 2013; Wheeler and von Braun 2013). Developing countries have argued that they need more resources to address food security, while developed countries have argued that they have already provided significant resources to address this issue. So, resource allocation is a contentious issue in the negotiation of *the Law of Food Security*. Therefore, there is a need to establish a mechanism for allocating resources that prioritize the needs of developing countries, such as through a global fund for food security. Similarly, technology transfer is another contentious issue in the negotiation of the Law of Food Security. Developing countries have argued that developed countries have been slow to share their technology and that they need more investment in research and development to develop their agricultural sectors. Developed countries have argued that they have invested significantly in research and development, and they should not be expected to share their technology for free. Therefore, there is a need to establish a framework for technology transfer that addresses the concerns of both developing and developed countries (King 2017; Jaykus et al. n.d.).

Although there are many contentious issues about the developed and developing, but there is a growing recognition among countries around the world that access to adequate and nutritious food is a fundamental human right. Therefore, there is a common ethic among countries to ensure food security for their populations (Frohlich 2019; Schanbacher 2010). At the national level, many countries have adopted policies and programs to address food security issues. For example, Brazil's Zero Hunger program has been successful in reducing hunger and malnutrition by providing food assistance, nutrition education, and support for small-scale farmers. Similarly, India's National Food Security Act aims to provide subsidized food to 75% of the rural and 50% of the urban population ("Commentary on the Indian Protection of Plant Varieties and Farmers' Rights Act 2001" 2016; Puri 2022). While the approaches to achieving food security may vary among countries, the underlying principle of ensuring access to adequate and nutritious food is a shared ethic. The growing recognition of the right to food as a basic human right reflects the common understanding among countries that food security is a fundamental aspect of human dignity and well-being ("Food Sovereignty, Food Security and the Right to Food" 2021). Therefore, *the Law of Food Security* can provide a platform for negotiating and implementing solutions to global food insecurity, promoting sustainable agriculture, and reducing poverty. The detail provision of the proposed *the Law of Food Security* is discussed in Chap. 6.

5.7 Conclusion

Analyzing the three major international food law, it is observed that while most of the international focuses on food safety standard during international food trade, they do not establish an operating global food system due to a lack of institutional framework. Food safety has been a leitmotif of international food policy in recent decades (Downes 2015) but focusing particularly on food safety cannot build food security. The presence of scattered international food law attempting to offer a mechanism to ensure food safety only means that there is currently no international food law governing the global food system. While contemporary international food law focuses on food safety and food trade, they resemble more to International Food Safety Law branch of International Food Law and cannot be referred to as International Food Law. However, they can be a part of International Food Law. Although food safety is pertinent to the development of food regulations, to build food security, it is important to form an institutional framework and governance. This leads to comprehend the essentiality of developing international food law. This dissertation proposes the development of international food law in a form of multilateral treaty. A global food governance model is proposed that addresses food security, food safety, responsible investment, accountability, monitoring, nutrition education, and an appropriate legal and policy framework to ensure equitable access to sufficient and nutritious food for everyone in the world. The subsequent Chap. 6 provides an outline for *the Law of Food Security* with essential provisions and the necessary of the proposed provision to establish a sustainable global food system.

References

Abbott KW (2017) Theorizing regulatory intermediaries: the RIT model. SSRN Electron J. https://doi.org/10.2139/ssrn.2925411

Abbott KW, Levi-faur D, Snidal D (2017) Theorizing regulatory intermediaries. Ann Am Acad Pol Soc Sci 670(1):14–35. https://doi.org/10.1177/0002716216688272

Alam S, Tomossy GF (2017) Overcoming the SPS concerns of the Bangladesh fisheries and aquaculture sector: from compliance to engagement. J Int Trade Law Policy 16(2). https://doi.org/10.1108/JITLP-01-2017-0002

Alemanno A (2015) The multilateral governance framework for food safety: a critical and normative overview. In: Food safety, market organization, trade and development. Springer International Publishing, pp 9–43. https://doi.org/10.1007/978-3-319-15227-1_2

Ayoki M (2018) Special and differential treatment of developing countries in the WTO agreement on trade facilitation: is there a cause for optimism? AJIS 8(4)

Bleicher SA (1969) The legal significance of re-citation of general assembly resolutions. Am J Int Law 63(3):444–478. https://doi.org/10.2307/2198866

Bronfenbrenner M (1956) Uses of agricultural surpluses to finance economic development in underdeveloped countries: a pilot study in India. Econometrica 24(3):361. https://doi.org/10.2307/1911643

Commentary on the Indian Protection of Plant Varieties and Farmers' Rights Act 2001 (2016) Farmers crop varieties and farmers rights. Routledge, pp 299–313. https://doi.org/10.4324/9781849775663-23

Das K (2008) Coping with SPS challenges in India: WTO and beyond. J Int Econ Law 11(4):971–1019. https://doi.org/10.1093/jiel/jgn033

de Chazournes LB (2009) Changing roles of international organizations: global administrative law and the interplay of legitimacies. Int Organ Law Rev 6(2):655. https://brill.com/view/journals/iolr/6/2/article-p655_14.xml [Accessed on 2023-09-20]

Disdier A-C, Fontagné L, Mimouni M (2008) The impact of regulations on agricultural trade: evidence from the SPS and TBT agreements. Am J Agric Econ 90(2):336–350. https://doi.org/10.1111/j.1467-8276.2007.01127.x

Downes C (2015) Worth shopping around? Defending regulatory autonomy under the SPS and TBT agreements. World Trade Rev 14(04):553–578. https://doi.org/10.1017/s1474745615000178

FAO (2020) The state of food security and nutrition in the world 2020. Food and Agriculture Organization of the United Nations

FAO (2021) The state of food and agriculture 2021. Making agrifood systems more resilient to shocks and stresses. Food and Agriculture Organization of the United Nations

Food sovereignty, food security and the right to food (2021) In: Handbook of critical Agrarian studies. Edward Elgar Publishing. https://doi.org/10.4337/9781788972468.00035

Frohlich X (2019) Food, politics and society: social theory and the modern food system. Soc Hist 44(4):520–522. https://doi.org/10.1080/03071022.2019.1655913

Gerber PJ (2013) Tackling climate change through livestock – a global assessment of emissions and mitigation opportunities. In: Most, vol 14, no 2. Food and Agriculture Organization of the United Nations (FAO), Rome

Hegde V, Wouters J (2021) Special and differential treatment under the World Trade Organization: a legal typology. J Int Econ Law 24(3). https://doi.org/10.1093/jiel/jgab033

Jaykus L-a, Acuff GR, Busta FF, Dickson JS, Hollingsworth A, Marcy J, Ann Marie M (n.d.) Managing food safety: a systematic approach. Food Technology Magazine. https://doi.org/10.1111/1750-3841.16532
Johnecheck W (2008) Trade in food: regulatory and judicial approaches in the EC and the WTO. By Alberto Alemanno. J Int Econ Law 11(2):495–500. https://doi.org/10.1093/jiel/jgn009
Kanu D, Oseni T (2022) United Nations and collective agreements on global security: issues and perspectives. J Contemp Int Relat Diplom 3(2):603–622. http://journals.abuad.edu.ng/index.php/jcird/article/view/65 [Accessed on 2023-09-20]
Keck A, Low P (2011) Special and differential treatment in the WTO: why, when and how? SSRN Electron J. https://doi.org/10.2139/ssrn.901629
King A (2017) Technology: the future of agriculture. Nature 544(7651):S21–S23. https://doi.org/10.1038/544s21a
Kurtz J (2007) A look behind the mirror: standardisation, institutions and the WTO SPS and TBT agreements. UNSW L J 30(2):504–523
Loibl G (2001) The role of international organisations in international law-making international environmental negotiations-an empirical study. Non-St Actors Int'l L 41(1). https://heinonline.org/hol-cgi-bin/get_pdf.cgi?handle=hein.journals/nonstata1&section=9
MacMaolain C (2008) The future control of food: a guide to international negotiations and rules on intellectual property, biodiversity and food security. Eur J Int Law 19(5). https://doi.org/10.1093/ejil/chn064
McGinnis MD (2011) An introduction to IAD and the language of the Ostrom workshop: a simple guide to a complex framework. Policy Stud J 39(1):169–183. https://doi.org/10.1111/j.1541-0072.2010.00401.x
Micheni A, Gicheru P, Kitonyo O (2022) Conservation agriculture for climate smart agriculture in smallholder farming systems in Kenya. In: Conservation agriculture in Africa: climate smart agricultural development (pp. 431–442). CABI. https://doi.org/10.1079/9781789245745.0027
Mingst K, Karns M, Lyon A (2022) The United Nations in the 21st century. Routledge.
Ostrom E, Gardner R, Walker J (1994) Rules, games, and common-pool resources. University of Michigan press.
Puri R (2022) India's National Food Security Act (NFSA). In: Food governance in India. Routledge India, pp 1–18. https://doi.org/10.4324/9781003272656-2
Renn O, Dreyer M (eds) (2009) Food safety governance. Springer, Berlin Heidelberg. https://doi.org/10.1007/978-3-540-69309-3
Rivera W (2005) Communication for rural development: challenge to diffuse development information on non-agricultural rural needs. J Int Agric Ext Educ 12(1). https://doi.org/10.5191/jiaee.2005.12108
Schanbacher WD (2010) The politics of food: the global conflict between food security and food sovereignty. Choice Reviews Online 47(12). https://doi.org/10.5860/choice.47-6944
Silverglade BA (2000) The WTO agreement on sanitary and phytosanitary measures: weakening food safety regulations to facilitate trade? Food Drug L J 55(4)
South Centre (2017) The WTO's special and differential treatment negotiations. South Centre 3
Surono A, Hidayati MN (2019) Special and differential treatment in the WTO agreement on agriculture and the benefits for developing countries. AJIS 8(4). https://doi.org/10.36941/ajis-2019-0046
Wheeler T, von Braun J (2013) Climate change impacts on global food security. Science 341(6145):508–513. https://doi.org/10.1126/science.1239402
Wilson J, Hufbauer CG, Kotschwar B (2001) Trade policy, standards, and development in Central America. The World Bank. https://doi.org/10.1596/1813-9450-2576
Wood R (2004) Recent developments at the Codex Alimentarius Commission and the inter-agency meeting with respect to food analysis. Accred Qual Assur 9(9):551–556. https://doi.org/10.1007/s00769-004-0835-9
Zhang Lili (2011) On logical justification and basic structure of the basic law of public Enterprise of China. US-China L Rev 8:332

Chapter 6
Introduction to the Law of Food Security: A Legal Framework for Global Food Justice and Sustainability

6.1 Introduction

Chapter 1 discussing the multidimensional aspect of food security, presented the persisting struggles the contemporary global food system is confronting. Conferring that the contemporary global food system is still observing food insecurity, even after 100 years of international food law, the chapter demonstrated the pressing need to analyze the contemporary international food law governing the contemporary global food system to comprehend and address the root of food insecurity. In this context, Chap. 2 focusing on the food insecurity, food security, and food safety makes an endeavor to comprehend the nexus between these three concepts argues that food security holds two aspects, first, in term of sufficient quantity of food (i.e. food insecurity), and second, in term of consumption of safe food (i.e. food safety) which is also similar in the context of food insecurity. So, food insecurity is the antithesis of the notion food security, however, food safety is a separate concept but cannot be represented or comprehended explicitly from both the concepts. Hence, food safety is an integral part of both the notion food insecurity and food security. Besides the concepts of food insecurity, food security, and food safety, Chapter 3 discusses an emerging and evolving concept i.e. food waste in the context of food security. The chapter asserted that food waste is exacerbating food insecurity. Chapter 4 analyzes 100 years of the development of international food law since the emergence of the concepts food insecurity and food security. Scrutinizing international food law, the chapter presenting the change in approach toward food insecurity and food security, argued that the contemporary international food law focuses more on food safety during international food trade by making resisting endeavor in harmonizing food safety standard. Holding these postulations in Chaps. 2, 3, and 4, Chap. 5 examines the major international food law and demonstrated that the main rationales for which food insecurity is still a global issue because, the major contemporary international food law addresses only food safety and lacks institutional framework to form food governance. Building on the foundation laid by the 1993 International Food Security

S. S. Silvee, *The Law of Food Security: The Legal Framework for a Sustainable and Equitable Global Food System*, LITES - Legal Issues in Transdisciplinary Environmental Studies 5, https://doi.org/10.1007/978-981-95-3636-8_6

Treaty (IFST), this chapter proposes a guideline for drafting a new multilateral treaty specifically focused on food security. i.e. Law of Food Security (*herein after referred as,* The Treaty).

This chapter as the major contribution addresses the findings of Chaps. 2, 3, 4, and 5, and endeavors to present *an outline of possible general provisions* for *the Law of Food Security* based on the persisting common grounds agreed by the both developing and developed countries. The 1993 proposed IFST aimed at guaranteeing the right to be free from hunger and the right to adequate food. The goal of the treaty is to obligate state parties to ensure food security for all, prioritizing the most food-insecure populations. The treaty is not an established international agreement as of the current knowledge cutoff date, but an advocacy initiative. The proposed IFST 1993, though unsuccessful, offers valuable insights into the potential for a more comprehensive and rights-based approach to global food security, paving the way for a more equitable and sustainable future. Given the urgency and complexity of the global food crisis, there is a need for innovative and transformative solutions that can address the root causes and systemic issues of food insecurity and foster a more resilient and inclusive food system.

The new Treaty aims to establish a global regime of food security governance based on the principles of human rights, environmental justice, social justice, and cosmopolitanism. This new Treaty is proposed as a response to the "shortcomings and gaps of the existing legal frameworks and institutions" that regulate the global food system. The Treaty seeks to create a binding and enforceable mechanism that can ensure the realization of the right to food for all people, protect the environment and natural resources that sustain food production, promote fair and equitable distribution of food and access to essential resources, and enhance global cooperation and solidarity in addressing the challenges of food security.

The outline of possible provisions for the Treaty contained in this chapter applies to the multi-factor decision-making schema to food security, highlighting the possible and imperative provisions to build a sustainable and equitable global food system, which will enhance the development of international food law. The outline is developed analyzing and considering the international documents (see Appendix A) and national food, agricultural, food, and nutritional policy and legislation of twenty (20) countries (see Appendix C) which prescribed both substantive and procedure measures. These twenty (20) countries include both developing countries, some of the countries dealing with food insecurity, and developed countries. Despite the effort to be as practical as possible, the use of this guidance document will require decision-makers to be knowledgeable in food security. The outline emphasizes on the importance and the need of the particular provision to avoid the presence of any provisions holding less imperative. Hence, the documents firstly legitimatize the need of the provision and secondly, delineate the possible features of the provision. This chapter examines the research question: What can be the cooperative outlines of rules to build a comprehensive international food law? What are the main barriers and risks for ratifying and implementing *the Law of Food Security* at the national, regional, and global levels?

The chapter is divided into five (5) main segments including introduction and conclusion. The second presents an overview of the international instruments. The third segment is divided into thirteen (13) segments discussing different provisions under *the Law of Food Security* which needs to implement at national level to contribute in building a sustainable and equitable global food system at international level through global food governance. As *the Law of Food Security* develops a model structure for a sustainable and equitable global food system so, the third segment analyzes the difficulties such a system construction will face in future practice.

6.2 Overview of the Relevant International Legal Instruments and Frameworks on Food Security and Sustainability

The promotion of food security and sustainability is supported by various international legal instruments and frameworks that recognize the right to food and the importance of sustainable agriculture and food system, which contribute to the discourse on food security. Some of the key instruments and frameworks includes the Universal Declaration of Human Rights (UDHR),[1] the International Covenant on Economic, Social and Cultural Rights (ICESCR),[2] and the Convention on the Rights of the Child (CRC).[3] However, these instruments lack clear enforcement mechanisms and guidelines for implementation. Similarly, the Paris Agreement on Climate Change[4] while crucial for mitigating climate change's threat to food production, doesn't directly address issues of food production, distribution, or access. The right to food is recognized as a fundamental human right by most of the key instruments and frameworks of United Nations, which requires a transformative approach to food systems that ensures its realization.[5] However, a food systems approach can help to better understand the complex and dynamic nature of the global food system and to identify opportunities for innovation and governance. But the existing legal frameworks and policies related to food security are often fragmented, inadequate, or

[1] UN Universal Declaration of Human Rights. 1948 Available at: www.ohchr.org/en/universal-declaration-of-human-rights. [Accessed on 2022-11-10]

[2] UN International Covenant on Economic, Social and Cultural Rights 1966 Available at: https://www.ohchr.org/en/instruments-mechanisms/instruments/international-covenant-economic-social-and-cultural-rights. [Accessed on 2020-12-10]

[3] UN. Convention on the Rights of the Child. 1989 Available at: https://www.unicef.org/child-rights-convention.[Accessed on 2022-11-10]

[4] United Nations Framework Convention on Climate Change (UNFCC) Paris Agreement 2015 Available at: https://unfccc.int/process-and-meetings/the-paris-agreement. [Accessed on 2022-11-10]

[5] Schutter, Olivier de. "Report/submitted by the Special Rapporteur on the Right to Food, Olivier de Schutter." (2011). Available at: https://www.ohchr.org/EN/Issues/Food/Pages/FoodIndex.aspx. [Accessed on 2022-11-10]

ineffective in addressing the root causes of food insecurity and supporting sustainable and equitable food systems.

In addition to these key instruments, several international frameworks and initiatives promote food security and sustainability, including the United Nations Sustainable Development Goals (SDGs),[6] particularly SDG 2 on "Zero Hunger," provide a comprehensive framework but are aspirational and lack strong enforcement mechanisms. Reports from expert bodies like the High Level Panel of Experts on Food Security and Nutrition (HLPE)[7] and the International Panel of Experts on Sustainable Food Systems (IPES-Food)[8] offer valuable insights, but their implementation relies on voluntary state action. These limitations expose the shortcomings of the existing legal and policy landscape. While recognizing the right to food and highlighting the importance of climate change mitigation, there remains a significant gap in terms of a comprehensive, enforceable framework that addresses the broader food system, promotes sustainability, and ensures accountability for achieving food security for all. However, these legal instruments and frameworks provide a foundation for the development of the proposed new *the Law of Food Security* and highlight the importance of international cooperation and political will in addressing the challenges of the global food system.

Despite the existence of international legal instruments and frameworks promoting food security and sustainability, significant gaps and limitations remain at the global level. One of the main limitations is the lack of binding and enforceable international agreements on food security and sustainability.[9] The existing legal instruments are largely non-binding and lack effective mechanisms for implementation and enforcement.[10] Furthermore, the global food system is complex, and existing legal frameworks often fail to account for the interlinkages between different sectors, including trade, environment, and human rights.[11] This can lead to conflicts and contradictions in the implementation of legal obligations. Additionally, there are unequal power dynamics between countries and regions, which can make it difficult to achieve consensus and implement effective policies.[12] Finally, there is a lack of effective monitoring and reporting mechanisms to ensure compliance with existing

[6] UN Sustainable Development Goals 2015 Available at: https://sdgs.un.org/goals. [Accessed on 2022-11-10]

[7] HLPE Agroecological and other innovative approaches for sustainable agriculture and food systems that enhance food security and nutrition, 2019.

[8] IPES-Food. Uniformity to diversity: a paradigm shift from industrial agriculture to diversified agroecological systems. International Panel of Experts on Sustainable Food Systems. 2016 Available at: https://www.ipes-food.org/_img/upload/files/UniformityToDiversity_F ULL.pdf. [Accessed on 2022-11-10]

[9] FAO. Legal and Institutional Frameworks for Food Security and Nutrition. Rome: FAO. 2018.

[10] IPES-Food. The New Science of Sustainable Food Systems: Overcoming Barriers to Food Systems Reform. Brussels: IPES-Food. 2017.

[11] Lang, Tim, and David Barling. "Food security and food sustainability: reformulating the debate." The Geographical Journal 178, no. 4 (2012): 313–326.

[12] S Orden, David. "Agricultural Policy, Agribusiness and Rent-Seeking Behaviour." (2011): 291–294.

legal frameworks and address emerging challenges in a timely manner.[13] Addressing these legal gaps and limitations is critical for achieving sustainable and equitable food systems and ensuring the right to food for all.

6.3 The Law of Food Security: Core Provisions and Objectives

This outline of the essential provision material under the Treaty has been prepared: firstly, for international organizations, national executive authorizes or government officials, and other "decision-makers" and appropriate stakeholders, who holds the responsibility for proposing, planning, developing, and implementing appropriate policies and decisions. The outline may be helpful to support food policymakers to influence those charged with food security, food safety, health, economic, and social policy development. However, the Treaty needs to be opened for signature, ratification, or accession under the umbrella of UN, its relevant Agencies, intergovernmental organizations, and other relevant organizations, such as FAOs, WHO, and WTO, for its effective implementation and the legal binding effect. The outline as a recommendation encourages international organizations to step forward in uniting countries to form legal cooperation to build a sustainable and equitable global food system by enabling a shared understanding of the food security and food governance. In pursuance of drafting The Treaty following fourteen (14) provisions mentioned below can be considered:

6.3.1 Basic Objective, Scope, and Sources

6.3.1.1 Basic Objectives

The objective of a law determines the outcome that the law intends to achieve and the scope of law delimitates the application of law. So, objective of a law desires for an operational plan prepared considering the barriers in application of the law to achieve the outcome of the law. Taking into account the interests of different stakeholders involved throughout the global food system, the Treaty is outlined in order to achieve the following objectives mention below, which is based on six (6) sustainable development goals (SDG): **Goal 2:** Zero Hunger, **Goal 3:** Good Health, **Goal 9:** Industry, Innovation and Infrastructure, **Goal 12:** Responsible Consumption and Production, **Goal 13**: Climate Action, and **Goal 17:** Partnerships for the Goals.

[13] UN Global Report on Food Crises. Rome: FAO. 2018.

(I) To implement food security in the human life cycle approach, by ensuring access to an adequate quantity of safe and nutritious food at affordable prices for people to live a healthy life.[14]
(II) To promote the production of food meeting safe and nutritious food standard that is harmonized and equitable.
(III) To build partnership in eradicating food insecurity, hunger and malnutrition.
(IV) To encourage development and implementation of national food security action plans, strategy and policy to ensure people's right to have safe and nutritious food for a healthy and active life and ensure rights of different stakeholders.
(V) To establish a sustainable and equitable global food system with legal cooperation and harmonization.

6.3.1.2 Scopes of The Treaty

Scope of The Treaty is fully extended to independent States and governing the interrelation of the States "in so far as States are the constituent units of the international legal order and exercise legislative, judicial, and custom-creating authority within it either directly or through international bodies to which they have entrusted a measure of such authority."[15] So, focusing on objectives mentioned in Sect. 6.3.1.1 the scope of the Treaty, including aspects of ensuring civil rights, human rights, economic and social rights, and a number of collective rights and obligations, as substantive requirements of the law, bestow definite obligations upon State Parties of the Treaty to acknowledge, respect, and maintain certain rights. As the scope of the Treaty governs the interrelation between the States in building food security and a sustainable and equitable global food system, so it must include provision to settle common conflict between the States.

6.3.1.3 Sources of The Treaty

According to Article 38 of the International Court of Justice (ICJ) Statute there are five major sources of international law, they are[16]:

(I) Treaties between States;
(II) Customary international law derived from the practice of States;
(III) General principles of law recognized by civilised nations; and, as subsidiary means for the determination of rules of international law; and
(IV) Judicial decisions and the writings of "the most highly qualified publicists".

Beside considering these fives sources as its sources, The Treaty also acknowledges article 53 of the Vienna Convention 1969 according to which a peremptory norm

---

[14] [India] The National Food Security Act [S] 2013.

[15] C. Wilfred Jenks. The Scope of International Law [J] British Year Book of International Law, 1954(31): 1–48.

[16] Statute of the International Court of Justice [S] 1946.

of general international law is one which is "accepted and recognized by the international community of States as a whole as a norm from which no derogation is permitted and which can be modified only by a subsequent norm of general international law having the same character." So, this article suggests, fundamental principles of the international legal order are not based on any special source of law or specific law-making procedure, instead States, as the authority on behalf of the international community, plays a special role in implementing the peremptory norm of international law, such as for the Law of Food Security. So, the obligations imposed upon the State Parties of The Treaty by peremptory norms affect the vital interests of the whole international community.

6.3.2 General Provisions or Definition of Basic Concepts Clauses

The general provisions or definition of basic concepts clauses is the building blocks of any law, as they a law is drafted by combining different, but relevant concepts together in harmony.[17] Often concepts are comprehended as words or rules holding interest to the law, however, concepts which functions as the building blocks of any law are different,[18] as it establishes basic operational imperatives to guide the outcomes of the law. Hence, to avoid any ambiguity within the law regarding the concepts, general provisional or definition of basic concepts clause is important. So, in pursuance of drafting general provisional clause under Food Treaty following basic concepts mentioned below can be considered which is defined for the purpose of the Treaty:

6.3.2.1 Accessibility of Food

According to the Committee on Economic, Social and Cultural Rights, "accessibility requires economic and physical access to food to be guaranteed. Economic accessibility means that food must be affordable. Individuals should be able to afford food for an adequate diet without compromising on any other basic needs, such as school fees, medicines or rent. Physical accessibility means that food should be accessible to all, including to the physically vulnerable, such as children, the sick, persons with disabilities or the elderly, for whom it may be difficult to go out to get food. Access

---

17 Ginsburg, Tom & Stephanopoulos Nicholas. The Concepts of Law [J]. The University of Chicago Law Review, 2017(84):147.

18 See, Giovanni Sartori. Concept Misinformation in Comparative Politics [J] Am Polit Sci Rev, 1970, Vol. 64: 1033, 1035; See also, Gary Goertz. Social Science Concepts: A User's Guide [M] Princeton University Press, 2006:27–67.

to food must also be guaranteed to people in remote areas and to victims of armed conflicts or natural disasters, as well as to prisoners."[19]

6.3.2.2 Adequacy of Food

The Committee on Economic, Social and Cultural Rights defined adequacy of food as "the food must satisfy dietary needs, taking into account the individual's age, living conditions, health, occupation, sex, etc. Additionally, food should be safe for human consumption and free from adverse substances, such as contaminants from industrial or agricultural processes, including residues from pesticides, hormones or veterinary, and should be culturally acceptable."[20]

6.3.2.3 Availability of Food

The Committee on Economic, Social and Cultural Rights, noted, "food should be available from natural resources either through the production of food, by cultivating land or animal husbandry, or through other ways of obtaining food, such as fishing, hunting or gathering […] it means that food should be available for sale in markets and shops."[21]

6.3.2.4 Affordability of Food

Affordability of food is although a key component of food security, however, the concept is not adequately defined which presents the need of an officially acceptable definition. However, The Treaty might consider defining the term as:

> Affordability of food as one of the key component of food security presents the economic aspect of food and society by conferring the nexus between price of the food, quantity of the safe and nutritious food daily consumed and the choice of food.

6.3.2.5 Carbon Neutrality

Carbon neutrality means "having a balance between emitting carbon and absorbing carbon from the atmosphere in carbon sinks. Removing carbon oxide from the atmosphere and then storing it is known as carbon sequestration. In order to achieve

[19] UNHCR, Fact Sheet No. 34: The Right to Adequate Food [S] 2010.

[20] Ibid.

[21] Ibid, at p.2.

net zero emissions, all worldwide greenhouse gas (GHG) emissions will have to be counterbalanced by carbon sequestration."[22]

6.3.2.6 Farmer

Farmer can be defined as legal entity or person carrying out the farming activity alone or in association with others, essentially to be agricultural (so growers, breeders, and conservers of plants), pastoral (so growers, breeders, and conservers of animals/fish), and forestry (so harvesters and conservers of forests). The Protection of Plant Varieties and Farmers' Right Act 2001 (India) defined famer as "any person who "cultivates crops by cultivating the land himself," "cultivates crops by directly supervising the cultivation of land through any other person," or "conserves and preserves, severally or jointly, with any person any wild species or traditional varieties or adds value to such wild species or traditional varieties through selection and identification of their useful properties."[23]

6.3.2.7 Food

According to U.S. Code, "food means (1) articles used for food or drink for man or other animals, (2) chewing gum, and (3) articles used for components of any such article."[24] Additional to this definition of food, the definition might consider including the "nutritional standard" of the articles or component collectively or individually forming food.

6.3.2.8 Food Insecurity

According to FAO, food insecurity "exists when people do not have adequate physical, social, and economic access to sufficient, safe, and nutritious food, which meets their dietary needs and food preferences for an active and healthy life."[25] Further FAO defined FIS as "a situation that exists when people lack secure access to sufficient amounts of safe and nutritious food for normal growth and development and active and healthy life."[26]

---

[22] European Parliament. What is carbon neutrality and how can it be achieved by 2050? [EB/OL] www.europarl.europa.eu/news/en/headlines/society/20190926STO62270/what-is-carbon-neutrality-and-how-can-it-be-achieved-by-2050 2019/2022-04-06.

[23] The Protection of Plant Varieties and Farmers' Right Act (INDIA), Section 2(k) [S] 2001.

[24] U.S. Code. Definitions; generally 21 321 [S] 1938.

[25] FAO. Trade Reforms and Food Security: Conceptualizing the Linkages. [R] Rome: FAO, 2003.; See also, Thomas, H. C. (eds.).Trade reforms and food security: country case studies and synthesis [M] Rome: FAO, 2006.

[26] FAO, IFAD. WFP, The State of Food Insecurity in the World 2013-The Multiple Dimensions of Food Security. [R] Rome: FAO, 2013.

6.3.2.9 Food Security

According to the CFS, "food security exists when all people, at all times, have physical, social and economic access to sufficient, safe and nutritious food to meet their dietary needs and food preferences for an active and healthy life."[27]

6.3.2.10 Food Safety

Food safety refers to the conditions and practices that preserve the quality of food to prevent contamination and food borne illnesses.[28] Further food safety is defined as, "protecting the food supply from microbial, chemical (i.e. rancidity, browning) and physical (i.e. drying out, infestation) hazards or contamination that may occur during all stages of food production and handling-growing, harvesting, processing, transporting, preparing, and disturbing."[29] So, "food safety is an assurance of the food against chemical, biological or physical conditions which may expose the user or consumer to a health hazard."[30]

6.3.2.11 Food Safety Standards

Adhering to the definition of food safety, food safety standards means a particular level or benchmark that food needs to attain pertaining to be safe for health prescribed under certain law.

6.3.2.12 Food Nutritional Standard

Food nutritional standards mean a particular level or benchmark that the nutritional value of food needs to attain by balancing the ratio of essential nutrients, such as, carbohydrates, fat, protein, minerals, and vitamins and other concerning nutrient requirements to prevent suffering from malnutrition and stunting.

---

[27] CFS. Reform of the Committee on World Food Security Final Version. CFS:2009/2 Rev.2. [S] 2009: 1.

[28] USDA.gov. What does food safety mean? [EB] http://ask.usda.gov/s/article/What-does-food-safety-mean

[29] Agrovoc.fao.org. Food Safety [EB/OL] https://agrovoc.fao.org/browse/agrovoc/en/page/c33995 2012/2022-04-06.

[30] Anon. Annual report [R] Ministry of Agriculture, 2007.

6.3.2.13 Food Governance

Food governance of food system refers "to a mechanism that facilitates debate, convergence of views and coordination of actions to improve food security at global but also at regional and national levels."[31] Renn and Dreyer (2009), in Food Safety Governance: Integrating Science, Precaution and Public Involvement, noted, food governance framework pertains to a set of challenges which includes: "(1) The demarcation and coordination between assessment and management of food safety threats; (2) The handling of scientific uncertainty; (3) The increase of transparency during the entire food safety governance process; (4) The involvement of a diversity of social groups and the wider public into the governance process; and (5) The handling of highly controversial food safety issues." Therefore, food governance involves many conflicting dimensions where international trade, environmental problems, food law, human rights, poverty reduction, and health problems are interconnected. It forms an institutional framework where multi-actors, such as, private and public actors, and multi-level actors, NGOs, and different layers of government works in cooperation with each other.

The first step in developing effective global food governance is to recognize the importance of local and global processes, both in terms of production and food access. On the production side, global food governance should involve coordinating investments in research, agronomy, and technology to ensure the production of safe and nutritious food. Additionally, incentives should be developed to encourage investment in sustainable practices, such as organic agriculture and conservation of natural resources. On the access side, global food governance should consider strategies for increasing food access to vulnerable populations and minimizing food waste. This should involve investments in infrastructure, such as storage and transportation, as well as legal reforms aimed at providing access to food for those living in poverty.

To ensure accountability and compliance, the developing of a global food governance model should include mechanisms for monitoring and evaluating implementation. Such mechanisms should also consider potential risks, such as climate change and its effect on food security, as well as the potential for exploitation of vulnerable populations in the process. Additionally, international treaties should be negotiated and signed to ensure adherence to agreed-upon standards. Compliance should also be assured through a robust regulatory framework, involving multiple governments, international organizations, and stakeholders.

In order to achieve the desired outcome of access to safe and nutritious food, a global food governance model should also involve consumer participation in decision-making. This should include initiatives to promote consumer awareness regarding food production and its effects on their health and the environment. It should also include mechanisms to increase consumer access to information and education, and ultimately, involve consultation with consumer groups prior to any policy decisions.

---

[31] FAO. Governance: The Right to Food [EB/OL] www.fao.org/right-to-food/areas-of-work/governance/en/ n.d./2022-04-06.

Seeing as global food governance is a complex undertaking, adequate resources must be allocated to ensure effective implementation. This involves collaboration between countries and other stakeholders to access funds necessary for research, development, and implementation of effective global food governance measures. It is also essential for representatives from all affected populations to be actively involved in the decision-making process, to ensure that the resulting model is equitable and serves the interests of all relevant parties. A successful global food governance model should take into consideration the local and global processes involved in the production, distribution, and access to safe and nutritious food. To this end, a robust legal framework, monitoring and evaluation mechanisms, and adequate resources must be established. Additionally, consumer involvement in decision-making should be encouraged, in order to ensure the necessary public engagement and accountability for the project.

6.3.2.14 Food Waste and Food Loss

Food waste is defined as "food appropriate for human consumption being discarded or left to spoil at consumer level – regardless of the cause."[32] And food losses "as a decrease, at all stages of the food chain prior to the consumer level, in mass, of food that was originally intended for human consumption, regardless of the cause."[33] So, food waste refers to edible and inedible parts of food eliminated, lost, or discarded from the food supply chain which as no longer required or intended for human consumption, regardless of the cause, decreases the quantity or quality of food in the global food system. Hence, food waste and food loss can be considered as interchangeable concepts.

6.3.2.15 Hunger

FAO defined hunger as "an uncomfortable or painful sensation caused by insufficient food energy consumption. Scientifically, hunger is referred to as food deprivation."[34] In UN report: Pandemic year marked by spike in world hunger, hunger is defined as "an uncomfortable or painful sensation caused by insufficient energy from diet. Food deprivation is not eating enough calories."[35] The report used the term hunger interchangeably with (chronic) undernourishment and measured by the prevalence of undernourishment. Here is noteworthy to mention according to FAO "all food insecure people are hungry but all hungry people are food insecure."

---

[32] FAO. Food losses and waste in the context of sustainable food systems [R] Rome: FAO, 2014.

[33] Ibid.

[34] FAO. Food Security Information for Action Practical Guides: An Introduction to the Basic Concepts of Food Security [R] Rome: FAO, 2008.

[35] WHO. UN report: Pandemic year marked by spike in world hunger [R]. Geneva: WHO, 2021.

6.3.2.16 Moderate Food Insecurity

In UN report: Pandemic year marked by spike in world hunger and according to FAO, moderate food insecurity is defined "a state of uncertainty about the ability to get food; a risk of skipping meals or seeing food run out; being forced to compromise on the nutritional quality and/or quantity of food consumed."[36]

6.3.2.17 Malnutrition

According to WHO, "malnutrition refers to deficiencies, excesses or imbalances in a person's intake of energy and/or nutrients. The term malnutrition covers 2 broad groups of conditions. One is 'undernutrition'—which includes stunting (low height for age), wasting (low weight for height), underweight (low weight for age) and micronutrient deficiencies or insufficiencies (a lack of important vitamins and minerals)."[37]

6.3.2.18 Severe Food Insecurity

According to WHO UN report: Pandemic year marked by spike in world hunger (2021) severe food insecurity is "running out of food; experienced hunger; at the most extreme, having to go without food for a day or more."

6.3.3 Fundamental Principles of the Law of Food Security: Legal Norms and Customary Principles

Legal norms and customary principles play an ethical and operational role as the basic principles of the law forming the basic foundation of the law and set a legal standard. So, legal norms and customary principles determine the rights, responsibilities, and duties of those who are the subjects of legal relations. Explicitly addressing these principles entangles the international obligations which exist as a part of international practices followed by multiple countries as general and consistent practice since long. So, a good portion of the chapter is dedicated to understand applicable and relevant "legal norms and customary principles" because of the significant role assigned to these principles in the theory of rights, responsibilities, and obligations. For the purpose of the Treaty the following eight (8) principles might be acknowledged by the Treaty, which will contribute in shaping the responsibilities and duties with the Treaty:

---

[36] Ibid.

[37] WHO. Malnutrition [EB/OL] www.who.int/news-room/questions-and-answers/item/malnutrition n.d./2022-04-06.

6.3.3.1 Common but Differentiated Responsibility

The Common but Differentiated Responsibility is a principle which traditional considered from environmental perspective. The principle is traditionally applied to form an outline of state responsibilities toward global environmental issues considering the history of global carbon contribution as a common global issue to maintain economic sustainability and development, and shared responsibility to reduce carbon emission based on the capacity to continue economic growth.[38] Although the principle has a significant history of use by the international community for global environmental issues, but gradually the principle is receiving increasing recognition in international law from other perspectives.[39] To comprehend the application of the principle in terms of food security, it is imperative to apprehend that the principle holds two terms, "common" and "differentiated responsibility." The term "common" presents "common concern of mankind" which "suggests that certain risks affect and are affected by every nation on earth".[40] So the term "common" includes any global issue that can affect the global community for which the global community should "cooperate in a spirit of global partnership."[41] On the other hand, the principle addressing the common global issues affecting the global community imposes responsibilities to solve the issue. However, the responsibilities are said to be "differentiated," where developed countries will carry a greater share of the burden than developing countries, so that not all countries contribute equally in terms of equity. So, application of the principle of common but differentiated responsibility contributes in policy development where to resolve a common global issue, such as food security, the global community work in cooperation with differentiated responsibilities. In terms of food security, this principle imposes differentiated responsibilities, where developed countries will carry a greater share of the burden than developing countries, to solve a common global issue. So, application of the principle of common but differentiated responsibility contributes in policy development where to resolve a common global issue, such as food security, the global community work in cooperation with differentiated responsibilities. The principle was acknowledged in BRIC's Joint Statement on Global Food Security, stating, "The developed and developing countries should address the food security issue according to the principle of common but differentiated responsibility. The developed countries should provide financial and technology support for developing countries in the field of food production capacity."[42] The Treaty can incorporate this principle by outlining

---

[38] Grubb, Michael.Planetary Economics: Energy, climate change and the three domains of sustainable development.1st ed [M] Routledge, 2014.

[39] Stone, C. D. Common but Differentiated Responsibilities in International Law [J]The American Journal of International Law, 2004, 98(2): 276.

[40] Ibid.

[41] The phrase appears in the Report of the World Summit on Sustainable Development, UN. Report of the World Summit on Sustainable Development, Johannesburg, South Africa. UN Doc. A/CONF.199/20 revised by UN Doc. A/CONF. 199/20/Corr. 1 [R] New York: UN, 2002.

[42] BRIC's Joint Statement on Global Food Security [EB/OL] brics.utoronto.ca/docs/090616-food-security.html 2009/2022-04-09.

differentiated obligations for developed and developing countries. Developed countries could contribute more resources toward technology transfer, capacity building, and financial assistance for developing countries to improve food production and distribution systems.

6.3.3.2 Food Sovereignty

According to the Declaration of Nyéléni in 2007, "Food sovereignty is the right of peoples to healthy and culturally appropriate food produced through ecologically sound and sustainable methods, and their right to define their own food and agriculture systems." In World Food Summit (WFS 1996), food sovereignty is stated to be "rooted in the ongoing global struggles over control of food, land, water, and livelihoods." So, food sovereignty encourages the development of framework and policy by putting producers and democracy at the center of the food system and recognizing food as a right and public good. Accordingly food sovereignty outlines the policy around seven (7) principles which: (1) Focuses on food for people; (2) Builds knowledge and skills; (3) Works with nature; (4) Values food providers; (5) Localizes food systems; (6) Puts control locally; and (7) Food is sacred.[43] The Treaty can respect food sovereignty by avoiding provisions that dictate national agricultural policies. Instead, it can focus on establishing a framework for cooperation and knowledge sharing, allowing states to choose appropriate food production and distribution methods based on their specific contexts.

6.3.3.3 Principle of Common Concern of Humankind

Dinah Shelton explaining the principle of common concern of humankind (Shelton 2017) defined the term "common concern" as "issues of common concern are those that inevitably transcend the boundaries of a single state and require collective action in response." Historically the principle is mostly used to address global climate change. The United Nations Framework Convention on Climate Change (UNFCCC) in 1992 acknowledged climate change as "common concern of humankind." Although "food insecurity" is not acknowledged as "common concern of humankind," but considering the number of hungry and malnutrition people according to latest reports, certainly "food insecurity" is a common concern of humankind which requires collective action in response. Moreover, the principle provides a framework where burden sharing approach is taken to resolve a global problem.

[43] SeedChange. What is food sovereignty? [EB/OL] https://weseedchange.org/food-sovereignty/ n.d./2022-04-09.

6.3.3.4 Principle of Equitable and Reasonable Utilization for Sustainable Use

This principle is the cornerstone and the fundamental doctrine of the UN Watercourses Convention encouraging utilization of international watercourse in an equitable and reasonable manner to provide the maximum benefit to each State.[44] However, this principle is equitably relevant to the Treaty. Considering the enormous amount of food waste throughout the food supply chain, addressing this principle will provide a framework to encourage equitable use of food resources, reasonable amount of food waste, and standard consumption of food for sustainable use. Here "sustainable use" reflects the need to balance economic and social values throughout the food supply chain and in consumption of food. So, the equitable and reasonable utilization might take into account below relevant factors and circumstances:

(I) The natural character, such as geographic, ecological, climatic, and other factor;
(II) Availability, accessibility, and affordability of safe and nutritious food among the population;
(III) Food demand among the population and nutritional level with population.
(IV) Effect of food trade among the States;

As stated in UN Watercourses Convention, "the weight to be given to each factor is to be determined by its importance in comparison with that of other relevant factors. In determining what is a reasonable and equitable use, all relevant factors are to be considered together and a conclusion reached on the basis of the whole."[45] The Treaty can incorporate this by establishing guidelines for equitable access to and sustainable management of resources essential for food production. It could encourage cooperation on water sharing, promote responsible fishing practices, and discourage unsustainable land-use practices. Similarly, in determining what is a reasonable and equitable use under the Treaty importance in comparison with that of other relevant factors should be given.

6.3.3.5 Precautionary Principle

Although precautionary principle is widely used from environmental perspective, however, the principle holds a broad philosophical and legal aspect; hence, its components are still evolving. According to Sands and Peel "the precautionary principle is an attempt to give the notion of precaution—understood as a form of addressing risk—legal status." Although the principle is widely used from environmental perspective, but its core element is to scientifically assess the presence of any potential threat or risk that can cause serious damage and develop action plan accordingly to mitigate the

[44] UN Watercourses Convention (UNWC) Fact Sheet No. 4 Equitable and Reasonable Utilization [S] 2006.

[45] Ibid.

adverse impact of the threat. Application of precautionary principle will contribute in developing an action plan to mitigate the adverse impact of any external threat on food security, such wartime, climate disaster, and so forth. The Treaty can integrate this by encouraging states to adopt precautionary approaches in food production. This could involve promoting practices that minimize the use of harmful pesticides and fertilizers, and encouraging research into sustainable alternatives.

6.3.3.6 Principle of Prevention

Principle of prevention is historical used in environmental context, but the core element of the principle is to assess any anticipated threat or risk that can cause serious damage and develop a long-time action plan accordingly to prevent or intercept the threat or risk. By acknowledging this principle will encourage proactive function and operation action plans within the Law of Food Security. For instance, according to World Bank (2018) "at present 10.8 percentages (approximately over 1.5 billion hectares) of world's surface holds arable land, i.e. land that can be used in crop production." According to Vincent and André (2018) gradually a percentage of land is degrading and loss of land in happening because of varies reasons such as climate change. Hence, it can be asserted that one of anticipated threat or risk that can cause serious damage to food security is loss of arable land. The principle of prevention by encouraging scientific assessment over the situation prescribes to propose an action plan to prevent and avoid any potential risk. So, principle of prevention promotes the development of coherent prevention policy to prevent or combat and avoid any potential risk by evaluating the potential risks and initiating collective protective measures for a reasonable period of time. Provided that, the prevention policy cannot be arbitrary, adversarial, or punitive in nature, it must be developed based on risk assessment, sufficient scientific evidence, and available pertinent information, and must bear a rational relationship. The Treaty can incorporate this by encouraging investment in resilient food systems less vulnerable to climate change and other environmental disruptions. This could involve promoting drought-resistant crops, diversification of agricultural practices, and early warning systems for climate shocks.

6.3.3.7 Principle of Harmonization

Principle of harmonization works on developing a consistent approach within the law by creating common standards, and uniform and coherent regulations. So, principle of harmonization encourages development of framework with common objective, strategy, and standard which requires joint and active cooperation to achieve a common goal. The Treaty can promote harmonization by establishing international standards for food safety and quality, while respecting the diversity of national food production systems.

6.3.3.8 Principle of Special and Differential Treatment (SDT)

The principle of SDT, considering the economic condition of the developing countries and to maintain equity within the global community, offers transition period to the developing countries with some limited flexibility to cope with the economic pressures. So, SDT principle offers a set of legal provisions and rights,[46] such as, offers support to developing countries to build the infrastructure by providing the right to the developing countries to ask for support and longer period of time to enjoy the support, obliges the developed countries to support the developing countries, implement reasonable technical standard, and so forth. This indicates SDT principle requires the developed countries to establish cooperation with developing countries by acknowledging the SDT principle, and document how they complied with SDT principle. The Treaty can integrate this by incorporating provisions for financial and technological assistance to developing countries to improve their agricultural infrastructure, access to resources, and food distribution systems. By incorporating these principles, the Treaty can establish a comprehensive framework for sustainable food security, one that promotes international cooperation, respects national circumstances, and encourages responsible management of resources essential for feeding a growing global population.

6.3.4 General Principles

General principles are often comprehended as legal norms and customary principles, and the thin difference is recurrently overlooked. Legal norms and customary principles form the basic foundation of the law and set a legal standard for determining the rights, responsibilities, and duties. Whereas, general principles are basic rules performing as subsidiary and integrative tools which outline the basis responsibilities and obligations necessary for the very functioning of a legal system with legal reasoning and fill actual or potential legal gaps[47] by adhering to legal norms and customary principles. According to Kirk (1994) general principles as basic rules offer discipline, set value, and form instrument. So, general principles help to develop an institutional framework and attributes to fill the gap by forming governance for a legal system by connecting concepts with objective and scope of the law to determine the responsibilities, duties, and power of the subjects or participants of the law. So, general principles are some propositions of law so fundamental that it will be found in virtually every legal system.[48] In pursuance of outlining general principles to develop

---

[46] L. McAuley. Special and Differential Treatment (SDT) in the World Trade Organization (WTO) [EB/OL] www.gtpalliance.com/trends-transforming-trade/special-and-differential-treatment-sdt-in-the-world-trade-organization-wto 2021/2022-02-20.

[47] Kohen, Marcelo, and Schramm, Bérénice. General Principles of Law [EB/OL] https://doi.org/10.1093/obo/9780199796953-0063 2019/2022-02-20.

[48] Apple, James G. What are General Principles of International Law? [J] The American Society of International Law and the International Judicial Academy, 2007. 2 (2).

an institutional framework forming governance under Food Treaty following can be considered:

Firstly, Principle 4 of the *FAO Principles* and Guidelines for National Food Control Systems (2013) can illustrate the roles and responsibilities of all participants to form food governance. It states:

1. 1. All participants in a national food control system should have specific roles and responsibilities clearly defined.
2. Food business operators have the primary role and responsibility for managing the food safety of their products and for complying with requirements relating to those aspects of food under their control.
3. The national government (and in some cases a competent authority) has the role and responsibility to establish and maintain up to date legal requirements. The competent authority has the responsibility to ensure the effective operation of the national food control system.
4. Consumers also have a role in managing food safety risks under their control and where relevant should be provided with information on how to achieve this.
5. Academics and scientific institutions have a role in contributing to a national food control system, as they are a source of expertise to support the risk based and scientific foundation of such a system.

Although food insecurity is a global issue, however, resolving the issue needs actions plans at national levels connected with the international level. Hence, the The Treaty as a multilateral agreement among the countries encourages the members or participants to establish National Food Control Systems, because the global food system as International Food Control System will only function effectively depending on the National Food Control Systems. Analyzing the countries profile in FAOLEX Database most of the countries have focused only on developing law aim at building food security in terms of consumption of food, such as, Australia, Bangladesh, Bhutan, Japan, Korea, Pakistan, and United Kingdom, and didn't put much attention on building food security in terms of quantity of food. However, there are countries like, Afghanistan, Canada, China, India, Netherlands, Singapore, and United States, who developed law addressing food security in terms of quantity of food. Here it is noteworthy to mention, many countries have development action plans but it does address the issue of food insecurity, Hence, secondly, considering the Principle 4 of the *FAO Principles* (2013), The Treaty encourages countries to establish mechanisms to monitor and evaluate the implementation of the treaty guidelines toward the progressive realization of the right to food, right to adequate food, right to safe food, farmers right, in the context of national food security control, in accordance with their capacity and by building on existing information systems and addressing information gaps. For, instance, considering the establishment of National Food Control Systems, United Republic of Tanzania's Food Security Department (FSD),[49] and Singapore's Food Security Agency (FSA)[50] can be consider developing following equivalent responsibilities and duties:

[49] Food Security Act [S] 1991.

[50] [Singapore] Food Agency Act 2019 [S] 2019.

1. "Advise the Government on Food Security policies considering the indicators to implement legal, policy and administrative measures, detect discriminatory practices and outcomes, and ascertain the extent of political and social participation in the process of realizing that right[51];
2. Co-ordinate food security policies being effected/implemented for the time being, by other bodies or organizations;
3. Implement or cause the implementation of different Government approved programmes on food security;
4. Initiate, undertake and participate in the collection, preparation and production of data on the food security for the use by the Government and its different bodies or organizations;
5. Monitor the country's food situation at all stages, that is soil preparation, planting growing, harvesting, storing, distributing and pricing and provide early warning service on food security;
6. Co-operate and exchange information about food security with different organizations at national regional and international levels;
7. Ensure the availability, within a required time, of relevant information and reports concerning food and agriculture generally;
8. Recommend to the Government the levels of exports and imports of major cereal staple crops annually; co-ordinate all activities concerning food aid;
9. Review and propose the nation's Strategic Grain Reserve level when required;
10. Procure crops for Strategic Reserve and to maintain the reserve;
11. Monitor the volume, type distribution and usage of farm inputs designed for food crops;
12. Prepare and implement the process of Strategic Grain Reserve recycling/rotation and release;
13. Participate and represent the Government in meetings and workshops concerning food security;"[52]
14. "To support regulation of the handling and supply of food to ensure that it is safe and suitable for human consumption and to promote public health;
15. To support regulation of businesses engaged in the handling or supply of food so as to minimize food safety risks and to ensure the provision of information relating to food to enable consumers to make informed choices;
16. To undertake or facilitate the education and training of persons engaged in the handling or supply of food to enable them to minimize food safety risks;
17. To support regulation of street hawking and trade fairs, whether or not engaged in the handling or supply of food;
18. To develop food education initiatives, including the publication of information to increase public awareness of food supply resilience, food safety, food standards and food labels;
19. To promote and develop the food industry, including any enterprise based on or using agri-biotechnology, agro-technology or food biotechnology;
20. To promote or undertake research into matters relating to food, the food industry, food businesses, food safety, food supply and on any matter which may be included in food standards;"[53]
21. To monitor and reviewing food industries and food security situation of vulnerable groups, especially women, children and the elderly, and their nutritional status, including the prevalence of micronutrient deficiencies.

---

[51] FAO. Voluntary Guidelines to support the progressive realization of the right to adequate food in the context of national food security. [R] Rome: FAO, 2005.

[52] Food Security Act [S] 1991.

[53] Singapore Food Agency Act 2019 [S] 2019.

22. To create public awareness among different stakeholders regarding food safety, food hygiene, food standard, packaging and labeling.
23. "To conduct 'Right to Food Impact Assessments' in order to identify the impact of domestic policies, programmes and projects on the progressive realization of the right to adequate food of the population at large and vulnerable groups in particular, and as a basis for the adoption of the necessary corrective measures."[54]

Considering the responsibilities and duties of FSD, other operational and functional departments such as Food Production and Control Department Food Safety and Nutritional Management Department (discussed in Sect.6.3.7 of this chapter), Anti-food waste Department (discussed in Sect. 6.3.8 of this chapter), Risk Management and Control Department (discussed in Sect. 6.2.9 of this chapter), and Knowledge Innovation and Management Department (discussed in Sect. 6.3.11 of this chapter), can be established. Beside, the responsibilities and duties general principles also as "gap filers" contribute by constructing and delimiting powers of the department.

6.3.5 Rights and Obligations

Rights are prerogatives that are exclusive privilege offered to a particular individual to have or do something, whereas, obligations are rudiment that is a course of legally bounded action one must do. So, rights and obligations play an important role in establishing a legal mechanism, as rights create obligations upon respective authority to build a mechanism to ensure rights. In pursuance of outlining the rights and obligations under Food Treaty, following can be considered:

6.3.5.1 Right to Adequate Food

"The right to adequate food is realized when every man, woman and child, alone or in community with others, has physical and economic access at all times to adequate food or means for its procurement."[55] An illustration of the right is given by United Nations Special Rapporteur, stating, "the right to have regular, permanent and free access, either directly or by means of financial purchases, to quantitatively and qualitatively adequate and sufficient food corresponding to the cultural traditions of the people to which the consumer belongs, and which ensures a physical and mental, individual and collective, fulfilling and dignified life free of fear."[56] According to Committee on Economic, Social and Cultural Rights (CESCR), available, accessible, and adequate are the core content of the right to adequate food.

---

[54] FAO. Voluntary Guidelines to support the progressive realization of the right to adequate food in the context of national food security. [R] Rome: FAO, 2005.

[55] CESCR. General Comment 12, Right to adequate food (Twentieth session, 1999) U.N. Doc. E/C.12/1999/5 [S] 1999.

[56] UNHCR, Fact Sheet No. 34: The Right to Adequate Food [S] 2010.

6.3.5.2 Right to Food

Right to food is evolving human rights sharing a strong nexus with right to life under Article 3 of UDHR; which obliges the government to ensure, firstly, life, food is the basic need of life, and secondly, security, food security is one of the essential security for living a health and active life. Beside, this Article 25 of UDHR also presents a direct connection by stating. "everyone has the right to a standard of living adequate for the health and well-being of himself and of his family, including food [...]." The right is also supported by article 11 of the ICESCR, Voluntary Guidelines to support the progressive realization of the right to adequate food in the context of national food security (Right to Food Guidelines, 2004), and Voluntary Guidelines on the Responsible Governance of Tenure of Land, Fisheries and Forests in the Context of National Food Security (2012). Right to food establishes three categories obligation: (1) The obligation to respect the right to food; (2) The obligation to protect the right to food; and (3) The obligation to fulfill the right to food in terms of available, accessible, adequate, and affordable.

6.3.5.3 Right to Have Safe Food

According to FAO, "food safety implies the absence or safe levels of contaminants, bacteria, naturally occurring toxins or any other substance that may make food injurious to health."[57] Moreover, the right to have food addresses food quality, safety standards, and food labeling. This right creates obligation on State Parties to take legislative and other measures needed to protect people, especially children, from taking unhealthy food by controlling advertising and promotions of unhealthy food, and marketing of food with misleading information.[58]

6.3.5.4 Right to Have Food of Own Preferences

Right to have food of own preference is a right of the consumer which is not much explored. However, the importance of food preference is explored by many scholars from food safety, food quality, food labeling, and legislative perspective.[59] Gradually over time the right is gathering importance and making its places in trade agreements. For instance, in United States—*measures concerning the importation,*

---

[57] FAO. Food: A fundamental human right. [EB/OL]. http://www.fao.org/focus/e/rightfood/right2.htm. n.d./2020-02-09.

[58] Luo, Y., & Wu, G. Food Safety Laws and Regulations [M] In Joseph J. Jen and Junshi Chen (eds.) Food Safety in China: Science, Technology, Management and Regulation. John Wiley, 2017: 345–361.

[59] See, Maidana-Eletti, Mariela. Promises and Perils of the TBT Agreement: Food Quality, Food Labeling and Market Access [J] The New Zealand Yearbook of International Law, 2014(12): pp. 65–76; See also, Skogstad, Grace. Internationalization, Democracy, and Food Safety Measures: The (Il) Legitimacy of Consumer Preferences?[J] Global Governance, 2001, 7(3): pp. 293–316.

marketing and sale of tuna and tuna products, the "first, demand equations for tuna products in the United States and Mexico are derived based on consumers' preference for yellow fin tuna and generic tuna (all other tuna) and consumption values in 2014."[60] Similarly, consumer preference also gained importance in Philippines—*Taxes on Distilled Spirits case.*[61] Although, in these cases "right to have food of own preference" is not acknowledged, but significant importance is given to "consumer preference" in terms of taste and rise in food prices[62] based on assessment of potential competition. However, the nexus between rise in food prices and "consumer preference" is well comprehended, demonstrating imposing significant costs on the purchase of food affects by "freezing consumer preferences" or "limiting consumer preferences," mostly because of the socio-economic condition of the consumer which is not argued in any cases. Hence, considering consumer tastes, habits, and socio-economic differ from one market to another, State Parties may consider ensuring right to have food of own preference which obliges the State Parties to maintain a decent food price—making food affordable to the maximum people.

6.3.5.5 Right to Adequate Nutrients

Right to adequate nutrients is an evolving food right based on the concept of "food and nutrition security." Although, the right holds historical ground from child's health perspective,[63] but not from adult's health perspective. In the mid-nineties, UNICEF working on children, particularly highlighted the importance of adequate nutrients for children for a healthy and active life evolved the understanding of undernourished and malnutrition health condition, and induced many international instruments.[64] For instance, Article 24 of the Convention on the Rights of the Child (1989) states that "States Parties recognize the right of the child to the enjoyment of the highest attainable standard of health" and shall take appropriate measures "to combat disease and malnutrition" through the provision of adequate nutritious foods, clean drinking water, and health care.

Malnutrition, including undernourished, can be in all forms diet-related non-communicable diseases such as wasting (too thin for height), stunting (too short for age), underweight, inadequate vitamins or minerals, overweight, obesity, and so forth. According to WHO, "1.9 billion adults are overweight or obese, while 462 million

---

[60] Appellate Body DS381—United States—Measures Concerning the Importation, Marketing and Sale of Tuna and Tuna Products WT/DS381/AB/RW/Add.1 [S] 2019.

[61] Appellate Body DS396—Philippines—Taxes on Distilled Spirits WT/DS396/15/Add.3 WT/DS403/15/Add.3 [S] 2009.

[62] Appellate Body Report.DS75: Korea—Taxes on Alcoholic Beverages [S] 1997; See also, Panel Report, DS87: Chile—Taxes on Alcoholic Beverages [S] 2000.

[63] Eide A, Eide W. Goonatilake S. Gussow J. Omawale (eds.) Food as human right [M] Tokyo: United Nations University, 1984; See also, Eide A. Right to adequate food as a human right [M] New York: United Nations, 1989.

[64] De Schutter, Olivier. The Right to Adequate Nutrition [J] Development, 2014, Vol. 57(2): 147–154.

are underweight. Globally in 2020, 149 million children under 5 were estimated to be stunted (too short for age), 45 million were estimated to be wasted (too thin for height), and 38.9 million were overweight or obese. Around 45% of deaths among children under 5 years of age are linked to undernutrition."[65] Hence, acknowledging right to adequate nutrients will bestow responsibility among the government to take appropriate measures to combat diseases relating to malnutrition and undernourished.

6.3.5.6 Farmer's Rights and Privileges

Article 9 of International Treaty on Plant Genetic Resources for Food and Agriculture (ITPGRFA) (2009) recognizes farmers' rights. As a means of recognizing the past, present, and future contributions of farmers in conserving, improving, and making available the plant genetic materials that are important for food and agriculture to build food security, ensuring farmers' rights and privileges is essential for the Law of Food Security. Attempting to address farmers' rights and privileges some countries have adopted specific legislation, such as the Protection of Plant Varieties and Farmers' Rights Act (2001) (India) and The Protection of New Plant Varieties Act of (2004) (Malaysia). So, in accordance with farmer's needs and priorities, each State Party should take appropriate measures to protect and promote farmers' Rights through legislation, including:

(I) "Protection of traditional knowledge relevant to plant genetic resources for food and agriculture;
(II) The right to equitably participate in sharing benefits arising from the utilization of plant genetic resources for food and agriculture; and
(III) The right to participate in making decisions, at the national level, on matters related to the conservation and sustainable use of plant genetic resources for food and agriculture".[66]
(IV) The right to equitably participate in famers' capacity building and training programs to bring innovation in food and agriculture.
(V) The right to land for farming and adequate market price.
(VI) The right to technological support and financial subsides.
(VII) Farmers' intellectual property rights.

It can be asserted that "farmers' rights must be viewed as collective rights and not as an individual right"[67] in a broader aspect which encourages the government of State Parties of The Treaty to enact legislation to safeguard farmers' right and privilege.[68]

---

[65] WHO. Fact sheets—Malnutrition [S] 2021.

[66] International Treaty on Plant Genetic Resources for Food and Agriculture [S] 2009.

[67] FAO. Stakeholders' Consultation on Farmers' Rights. In African Position Paper [R] Rome: FAO, 2016.

[68] Adhikari K, et al. What Should Farmers' Rights Look Like? [J] The Possible Substance of a Right. Agronomy, 2021, 11(2):367.

6.3.6 *Provision on Establishing Carbon Neutrality in Global Food Supply Chain*

Food as a biological necessity forms an instrumental relationship which intricate by nature. It is well agreed by the scientific community that food is the only source of energy for the living organism, which helps a living body to build, re-build, grow, repair, and maintain. However, food can also have negative consequences on a living body, such as obesity, arthritis, diabetes, and heart disease, mostly because of the way food is produced and consumed. The distinct method of food production and different way of consumption has created an innovative and diversified relationship that is economic nature. In addition, as a source of social identity food plays one of the significant roles by forming a symbolic relation holding political characteristics. Colás et al. rightly addressed, "[W]e are [...] no longer just what we eat, but also what and how we consume." At present food does not only play a significant role in health context but also plays a significant role in economy, social, cultural, and historical context (Alinovi et al. 2008).

Taking a close look at food as a biological necessity, to create energy required for any a living organism to function, three kinds of food consumption habits are observed among the living organism—Herbivores, Carnivores, and Omnivores. Herbivores rely only on plant materials; in contrast, carnivores rely only on meat. On the other hand, omnivores prefer to eat both meat and plant matter. This indicates food for the living organisms are derived from living organism; quite like cell repairing cell. Focusing on microbiology of food McGee (1997) suggests food when decomposed releases "water molecules, carbohydrates, proteins, and fats." These carbohydrates along with fat acts as the major sources of energy that a living body needs to function and protein helps in biological processes. Therefore, it won't be imprecise to assert food holds a physical, chemical, and biological characteristic that contributes in a number of biological systems in a living body and helps the body to carry out specific functions necessary for everyday living; making food consumption compulsory for living organism and creating a positive instrumental relationship. However, the relation gets intricate in nature when it comes to human, mostly, because of the consumption habits leading to negative consequences such as obesity, heart disease, diabetes, and so forth.

The modern way of food consumption has developed over time. However, the Industrial Revolution has forever changed the food system from food production to consumption and the world has seen different manufactured food commodity, for instance, Coca-Cola in the late nineteenth century. This change gave "an entirely new style of commodity in the form of manufactured foods."[69] The modern food system suggests, "industrialization plays a central part in the commoditization of food in the transition to capitalist society [...]the food system comes to be considered a means of profit."[70] Therefore, commoditization of food has given food an economic

[69] Hanson, V. Davis. Fields without Dreams: Defending the Agrarian Idea [M] New York: Free Press, 1996.

[70] Colás, A. et al. Food, Politics and Society [M] California: University of California Press, 2018.

characteristic apart from its physical, chemical, and biological characteristics. The transformation of the food system, means, the way food is produced, distributed, and consumed, has shaped economy as well as affected the environment, in terms of greenhouse gas emission. So, commodification of food is not only promoting economic industrialization driven by changing technology, but also inducing adverse environmental impact.

Food is depended on climate, and the change in climate is the result of greenhouse gas (GHG) emission. According to reports, food is responsible for approximately 26% of global GHG emissions, livestock & fisheries account for 31%, crop production accounts for 27% agriculture and other land use for food emits 24%, supply chains account for 18%, and our way of food consumption emits 25.0% GHG.[71] So, as a whole the global food system, agriculture to manufactured food along with food consumption and waste disposal, is one of the key contributor to GHG emission causing climate change. On the other hand, "climate change threatens food supply through increasing temperatures, changing precipitation patterns, and greater frequency of some extreme events"[72]; as climate change constraints food production, eventually it affects food supply and food price. So, it is apprehended that if the contemporary food production and food commoditization activities holds a negative impact on climate change, simultaneously, climate change holds a negative impact on the global food system by threatening food supply, in terms of food production, food insecurity, availability, access, utilization, and stability.

In this context, the Treaty aims at reforming the global food supply to build a sustainable global food system encourages the State Parties to implement the principle of "Double Carbon" which offers a legal mechanism by developing a set of legal provisions and rights considering the balance between carbon peak and carbon neutrality. The principle of "Double Carbon" offers a dual policy which encourages building an economic system of green, low-carbon, and circular development, and the energy utilization efficiency of key industries.[73]

The principle of Double Carbon—dual policy is an emerging concept of the new development policy adhered by China. Adhering to the principle China developed its action plan along with guidelines forming a mechanism.[74] According to the China's double carbon –dual policy "the efforts to reduce pollution and carbon emissions must be balanced with the need to ensure the security of energy, industrial chains, supply chains, and food, as well as normal daily life. We need to respond appropriately to

---

[71] H. Ritchie. Food production is responsible for one-quarter of the worlds greenhouse gas emissions [DB/OL] 2019. https://ourworldindata.org/food-ghg-emissions 2022-02-24.

[72] Shukla, P., Skeg, J., Buendıa, E., et al. Climate Change and Land: an IPCC special report on climate change, desertification, land degradation, sustainable land management, food security, and greenhouse gas fluxes in terrestrial ecosystems. [R] IPCC, 2019.

[73] CGTN. China should proceed from reality to achieve 'double carbon' goal: official [N] CGTN. 2021-10-21.

[74] Working Guidance for Carbon Dioxide Peaking and Carbon Neutrality in Full and Faithful Implementation of the New Development Philosophy [EB/OL] https://en.ndrc.gov.cn/policies/202110/t20211024_1300725.html 2021-10/2022-04-13

any economic, financial, and social risks that may arise during the green and low-carbon transformation to prevent any excessive response and ensure carbon emissions are reduced in a safe and secure way." Promoting double carbon—dual policy, the following guidelines can be considered under the Law of Food Security, which can be implemented by Food Production and Control department established under FSD at national level through action plans:

1. Optimizing and upgrading industrial structures by moving faster to promote green agricultural development and improve carbon sequestration and efficiency in agriculture
2. Firmly curbing irrational expansion of energy-intensive and high-emission projects by introducing production capacity control policies.
3. Vigorously developing green and low-carbon industries by establishing green manufacturing system and embed the internet, big data, artificial intelligence, 5G, and other emerging technologies into green and low-carbon food supply chain.
4. Improving the transportation structure vigorously by developing multimodal transportation and raise the proportion of rail and water transportation to achieve a sustained reduction in energy consumption and the CO_2 emissions intensity in the food transportation.
5. Encouraging the use of energy-conserving and low-carbon transportation vehicles to promote intelligent transportation in food transportation.
6. Strengthening basic research and research on cutting-edge technologies to formulate an action plan to ensure science and technology better support endeavor to peak carbon dioxide emissions and achieve carbon neutrality and develop a technological roadmap to carbon neutrality in food industry or food supply.
7. Speeding up the research, development and dissemination of advanced and applicable technologies.
8. Accelerating the development of a green trade system to promote food trading.
9. Strengthening international exchanges and cooperation to promote the establishment of a fair and rational system for global food governance based on mutually beneficial cooperation.
10. Refining standard and measurement systems for carbon dioxide peaking and carbon neutrality in food supply chain.
11. Enhancing statistical and monitoring capacity for energy consumption throughout the food supply chain.

6.3.7 Provision on Establishing Food Safety Standard and Nutrition Standard

Establishing food safety and nutrition standard under the Treaty gives the treaty its comprehensive nature. As the Treaty focuses not only on food security in terms of quantity of food, but also, on food security in terms of consumption of safe, hygienic, and nutritious food. In order to establish food safety and nutrition standard under the Treaty references can be acquired from the experiences of SPS Agreement, TBT Agreement, CODEX, and FAO Guidelines. So, The Treaty bestows upon the State Parties the right to take sanitary and technical measures necessary for the protection of human life or health and determine the standard of protection, provided that,

1. "measures be applied only to the extent necessary to protect human life or health; based on scientific principles and sufficient scientific evidence"[75];
2. "bearing a rational relationship to the measure, be sufficient to demonstrate the extent of the risk which the measure is supposed to address, and be of the kind necessary for a risk assessment"[76];
3. The risk assessment must establish with sufficient scientific evidence 'irreversible risks', e.g. life terminating, damage to human health and so forth.[77]
4. The characteristics of the measures should be reasonable, best practiced, logical, standard and harmonized, bearing a rational relationship between the quality and quantity of the scientific evidence provided by the risk assessment[78]; and not inconsistent with the basic principle of the Treaty.

The Treaty does not encourage establishment of measures based on "appropriate level of protection." Rather it encourages, the standard of protection should be reasonable based on scientific evidence of evidence *irreversible risks*. Furthermore, for establishing nutrition standard, the Treaty encourages the State Parties to develop "nutrition guidelines and standards"—forming a framework and implementation of action plan. Development of nutrition guidelines and standards was recommended in UNSCN ICN2 Framework for Action at Recommendation 16, stating:

"Establish food or nutrient-based standards to make healthy diets and safe drinking water accessible in public facilities such as hospitals, childcare facilities, workplaces, universities, schools, food and catering services, government offices and prisons, and encourage the establishment of facilities for breastfeeding."[79] Similarly, WHO in Global action plan for the prevention and control of non-communicable diseases 2013–2020 also stated, "Promote the provision and availability of healthy food in all public institutions including schools, other educational institutions and the workplace. For example, through nutrition standards for public sector catering establishments and use of government contracts for food purchasing."[80]

---

75 Panel Report, DS292—European Union (formerly EC)—Measures Affecting the Approval and Marketing of Biotech Products [S] 2006.

76 Panel Report, DS392—United States—Certain Measures Affecting Imports of Poultry from China [S] 2010.

77 Appellate Body Report, DS26—European Union (formerly EC)—Measures Concerning Meat and Meat Products (Hormones) [S] 1998.

78 Appellate Body Report, DS76—Japan—Measures Affecting Agricultural Products II [S] 1997.

79 FAO & WHO. 2014. Conference outcome document: Framework for action [C] Second International Conference on Nutrition, Rome, 19–21 November 2014. Rome, Italy.

80 WHO. Global action plan for the prevention and control of noncommunicable diseases 2013–2020. [R] Switzerland: WHO, 2013.

At present many countries like Afghanistan, Bangladesh, Canada, China, India, Japan, Nigeria, United Kingdom, United States, and so forth have developed "Dietary Guideline," however, provide advice on what to eat and drink to meet nutrient needs to promote good health, and prevent disease. So as a dietary guideline advices the people what to eat and drink to meet nutrient needs, it does not perform as a nutrition guidelines and standards. Hence, the dietary guideline does not provide an adequate guideline implementation and monitoring system. Furthermore, it is observed the dietary guideline does not provide any ground level monitoring system to assess the nutritional value of food. Assessment of nutritional value of food is important to ensure that the adequate quantity of nutrition is acquired from the food. According to study, "because of soil depletion, crops grown decades ago were much richer in vitamins and minerals than the varieties most of us get today."[81] The study also addressed other similar studies: Kushi Institute analysis of nutrient data from 1975 to 1997 found that average calcium levels in 12 fresh vegetables dropped 27 percent; iron levels 37 percent; vitamin A levels 21 percent, and vitamin C levels 30 percent. Study of British nutrient data from 1930 to 1980, published in the *British Food Journal*, found that in 20 vegetables the average calcium content had declined 19 percent; iron 22 percent; and potassium 14 percent. These studies show overtime because of soil depletion nutrients value of grown crops gradually decreasing. Hence, to ensure the nutritional value of food is at a consistent and standard level framework and implementation of action plan is essential. The Treaty encourages the State Parties to develop a "nutrition guidelines and standards" not a dietary guideline. The "nutrition guidelines and standards" shall promote the development of a framework and mechanism to ensure nutritional value of food.

6.3.8 Provision on Anti-food Waste Measures

Considering the findings of the study in Chapter Three which substantiated that food waste exacerbates food security, exhausts resource, affects sustainable development and the world adversely, and holds adverse effect on social atmosphere; and that food waste occurs across the various stages of global food supply from farm level to processing and distribution, and at consumer level, the study recommends development of an anti-food waste system. Food waste is a not only a global issue, it's a problem which occurs at national level but makes a global impact. So, without addressing the problems at national levels, food waste at global level cannot be resolved. The prerequisite of developing an anti-food waste system is legislation, as legislation drives standards which ultimately determine practices, procedures, responsibilities, rights, remedies, and develops an institutional framework. Hence, the study recommends provision for "Anti-food waste measures" within The Treaty,

---

[81] Scheer, Roddy, and Dough Moss. Dirt Poor: Have Fruits and Vegetables Become Less Nutritious? [EB/OL] www.scientificamerican.com/article/soil-depletion-and-nutrition-loss/ 2011/2022-04-13.

which will form a legal cooperation at international level and will develop a harmonized anti-food waste system at national level, that will, collectively, reduce food waste at global level and contribute in building food security. At present the international instruments governing the contemporary global food system hold no anti-food waste provisions; and focus more on maintaining food standards and ensuring food safety during food trade. For instance, SPS Agreement, which forms a legal foundation for ensuring food safety in international food trade. SPS Agreement concerns the application of basic rules for food safety and animal and plant health regulations. As SPS Agreement permits governments to determine and "maintain an appropriate level of sanitary and phytosanitary protection," provided that the measure is not more trade restrictive, governments hold the privilege to determine their own "appropriate level of food standards."[82] Although, the agreement obliges that the measures to ensure food safety should be based as far as possible on the analysis and assessment of objective scientific data. However, it has also provided the privilege to set a higher level of protection different from the international standard.[83] Hence, enormous amount of food is wasted to produce food that matches this appropriate level of food standard. Furthermore, these international food standards are developed focusing on food safety, not acknowledging the enormous amount of food waste it can create. So a balance between food safety standards and food waste is important. So, in pursuance of drafting "Anti-food Waste Measures" within the Treaty below aspects can be considered.

Beginning with the basic structure of the "Anti-food Waste Measures." Creating legally binding obligations by setting up a legal and institutional foundation for anti-food waste system, the Treaty must be aimed at:

1. Reducing food waste across the global food chain without compromising with the general food safety and standards that must be meet to be suitable for human consumption;
2. Ensuring global food security by ensuring national food security;
3. Conserving resources;
4. Protecting the environment; and
5. Promoting sustainable economic and social development.

Considering these, the Treaty should provide legal and appropriate definitions of different terms used to avoid ambiguity and bring clarity, such as food, food waste, different stages of the global food chain, and so forth. Besides, addressing the different stages of the global food chain, the Treaty should provide guideline to establish national anti-food waste system by –

(I) Setting up anti-food waste tasks targeting different stages of the global food chain;
(II) Setting up food waste supervision and management unit;
(III) Setting up market supervision and management;
(IV) Establishing anti-food waste norms applicable in all food sector;

---

[82] SPS Agreement [S] 1995.

[83] Ibid.

(V) Adopting technically feasible and economically reasonable measures;
(VI) Adopting monitoring and investigating;
(VII) Setting up grain management, loss reduction, transportation, processing and storage mechanism;
(VIII) National subside system to develop food waste control mechanism;
(IX) Analyzing and evaluating national food waste situation; and
(X) Publishing annual anti-food waste situation of the country.

Moreover, the Treaty should oblige governments to develop national anti-food waste system by promoting the development of working plan focusing on food conservation and waste reduction, reducing food waste in the food and catering services, promoting rational use of food, enforcement and implementation of law, and strengthening organizational supervision and management.

Additionally, considering the principle of common but differentiated responsibilities, the Treaty creating legally binding obligations upon the countries to implement national anti-food waste legislations to prevent and reduce food waste at different stages of the food supply chain, such as at production level, food service level, consumers level, i.e. household, retailers level, and so forth. This will set up legal obligations and responsibilities upon food producers, food traders, retailers, and consumers to take measures to reduce food waste. By these, the Agreement can provide a blueprint to reduce food waste at national level. Because when food waste at national level is reduced, it will eventually reduce the food waste at international level.

Few countries like, China,[84] France,[85] South Korea,[86] Italy,[87] Australia,[88] and so forth have undertaken initiatives to reduce food waste by promoting "National Anti-food Waste Work Plan" and law. Here it is noteworthy to mention that China's "National Anti-food Waste Work Plan"[89] has drafted and works lightly different from other countries as it have given major important to food waste in food service industry at consumer level, i.e. canteen service, dinning management catering service, and so forth. Notwithstanding, the plan also focused on other stages of food supply chain. These policies and strategies undertaken by different governments to some extend have reduced food waste at national level, however, didn't contributed much in building food security at international level; mostly because these strategies couldn't halve food waste significantly. However, the success of the China's "National Anti-food Waste Work Plan" is yet to measure as the plan and the law came into enforce very recently. To make a significant reduction in food waste cooperation between

[84] [China] Anti-food waste work plan [S] 2022.

[85] [France]LOI n 2016-138 du 11 fvrier 2016 relative la lutte contre le gaspillage alimentaire (1) [S] 2016.

[86] D. Steffen. South Korea Has Almost Zero Food Waste, Here's How [EB/OL]. www.intelligentliving.co/south-korea-zero-food-waste/ 2019/2022-02-30.

[87] [Italy] Provisions concerning the donation and distribution of products food and pharmaceuticals for purposes of social solidarity and for the limitation of waste. Gazzetta Ufficiale [S] 2016.

[88] [Australia] National Food Waste Strategy [S] 2017.

[89] Anti-food waste work plan [S] 2022.

developed and developing countries is prerequisite also with global anti-food waste system.

It is imperative to develop national anti-food waste system to reduce global food waste as food waste is mostly generated in the production level, followed by the processing and storage, then consumer level, and retailers' level. For instance, to reduce food waste at production level i.e. agricultural production level, because of weather and pests, knowledge sharing and technological innovation is important, which requires cooperation between developed and developing countries and agricultural subsidies. Similarly, lack of adequate food storage facilities often generates food waste. Hence, it is important that the Treaty encourages and develops legal cooperation among countries to reduce food waste across the global food supply chain. In doing so, the Treaty must acknowledge the principle of "Special and Differential Treatment" to encourage developed countries or importing countries to investment and technological support the developing countries or exporting countries to establish and develop national anti-food waste system which will eventually contribute in global food security; as if food waste is reduced, surplus food will be produced for international food trade, which will eventually bring economic development. So, the Treaty can encourage reciprocity and non-reciprocity agreements between countries. Furthermore, the Agreement must also oblige the food-importing country to set up national food standards and food safety level which is technically feasible and economically reasonable measures and does not generate food waste in food exporting country. Any such food safety and food standard that generates enormous food waste above the neutral level of "Food Waste Severity Measuring Scale" (Fig. 3.1) must be reconsidered with evidence-based risk assessment and scientific evidences. Moreover, food safety and food standard measures by food-importing country that generates enormous food waste above the neutral level of "Food Waste Severity Measuring Scale" (Fig. 3.1) must be considered as "transnational harm," both in terms of economy and national food security. Additionally, encouraging multiple measures, precise policy, scientific management, and social co-governance, The Treaty should advocate healthy, green, and low-carbon lifestyle, resource-saving, and environmentally friendly consumption methods.

6.3.9 Establishing Risk Management and Control System

Risk management measures develop risk management and control system which is an important process because as it prescribes the development of necessary tools to adequately identify, analyze, and prevent the potential risks along with prescribing necessary precautionary and mitigating measures to prevent the adverse impact of any potential risks. In order to comprehend the essentiality of risk management and control system the ongoing global financial crisis of 2007–2009, Covid-19 situation, and Russian Federation and Ukraine crisis 2022 can be considered.

The relation between financial crisis and food security is apprehended by studying the global financial crisis of 2008–2009.[90] During the food crisis of 2007–2008, restrictive food trade law was introduced to insulate domestic food prices from the international price surge.[91] However, such protectionist measures (e.g. export and import barriers) ended up leading to even higher food prices, aggravating food insecurity in net importer countries and inflicting long-term damage on the international food trading system. Consequently, the prices of food, in particular rice, wheat, and maize, increased substantially globally which increased undernourished people, food insecurity, 50 million people were made poor in 2008 because of high food prices, created civil unrest in many countries and so forth.[92]

Covid-19 outbreak happened in Wuhan, China in the late December 2019 which is considered to be limited within China till January until United Kingdom reported its first case in February 2020. Since then, the outbreak of Covid-19 has gradually spread worldwide and within 100 days the virus spread to nearly 169 countries[93] which now increased to 221 countries. In the year 2020 Covid-19 pandemic attributed nearly 3.3 million reported global deaths, and more than 1.2 million deaths were officially unreported[94]; establishing its widespread nature and global impact. These suggest February 2020 can be considered as the starting time of the Covid-19 era as it gradually started affecting globally. However, the pandemic situation is still increscent, but the severity of the situation started descending after the global administration of vaccination which begun in late January 2021. And the world is moving with the "Covid-19 normal."[95] Hence, it is difficult to predict the end time of Covid-19 era. As the pandemic situation is holding for a long time, it won't be imprecise to address the year 2020 as the beginning of "Covid-19 era or global pandemic era"—mostly, because the world has witnessed global pandemic situation for the first time. Although, later many countries in the year 2021 have unclench the Covid-19 policies, but the Covid-19 era still prevails.

Since, its gradual spread the Covid-19 situation has influenced different strategies and policies, such as, lockdown or closure, travel restrictions, maintaining social distancing, specific health system, and so forth which are globally implemented at different levels. Although Covid-19 policies are implemented differently in different

---

[90] Anderson, K., Ivanic, M., Martin, W. Food Price Spikes, Price Insulation, and Poverty [M] In J.-P. Chavas, D. Hummels, B. Wright, (eds.) The Economics of Food Price Volatility. Univ. Chicago Press, 2014.

[91] Kijin Kim, Sunae Kim, and Cyn-Young Park. Food security in Asia and the Pacific amid the COVID-19 pandemic [J] ADB Briefs 2020 (139).

[92] FAO, High food prices: the food security crisis of 2007–2008 and recent food price increases—Facts and lessons [R] Rome: FAO, 2009.

[93] S. Clarke. How coronavirus spread across the globe -visualised [N] The Guardian 2020-04-09.

[94] WHO. The true death toll of COVID-19: estimating global excess mortality [EB/OL]. www.who.int/data/stories/the-true-death-toll-of-covid-19-estimating-global-excess-mortality 2020/2021-10-09

[95] The term is used to indicate the covid-19 prevention and control measures, such as health measures, travel restrictions and so forth, becoming the part of standard lifestyle.

regions and countries at different times, however, some common Covid-19 policies were implemented globally since the outbreak. In the year 2020, the Covid-19 has shown evolution with increasing number of infections and death. Furthermore, Covid-19 control and prevention measures are one of the features of the timeline that progressively become the new normal. Later the development of vaccination and advancement of the treatment along with the health system became among the other developed aspects. These unique features, policies, and developments contribute as the distinctive character of the Covid-19 era, which slowly and steadily affected all the four pillars of food security.

Illustrating how Covid-19 crisis affects food security, we start the discussion with food availability and stability, which is important to examine first from the food supply chain perspective and then economic perceptive. One of the main causes that contribute in food insecurity is sharp declines in food production. Covid-19 crisis is impacting the agricultural food production steadily and the impact varies differently by product and region. As agriculture food production plays a role in manufactured food production so any affect in the agriculture food production has an impact on manufactured food production also on food price. The Covid-19 lockdown or closure policy has impacted the food production line more intensely, because food industry produces daily life-essential products.[96] So, if food distributors and processors are unable to produce and distribute food at an affordable price, food security is at risk. Although, according to Organisation for Economic Co-operation and Development (OECD) "risk associated with food security isn't related to food accessibility, [rather] it's identified with shoppers' approach to food."[97] However, according to OECD "buyers sometimes experienced empty racks in shops during the start of the epidemic,"[98] which indicates the impact of Covid-19 crisis on food supply; mostly because shortage or limitation of labor due to lockdown, travel restriction, and health measures. So, the impact of shortage of labor on food production can be comprehended and presumed. Furthermore, as food industry is associated with various industries, hence, temporarily shut down of associated businesses had an impact on food supply in term of quantity of available food.

Additionally, Covid-19 crisis has also impacted in terms of consumption of safe food. For instance, cold chain system, "a specialized department of the food system supply chain, aims to guarantee food security and reduce the food waste caused by changes in physicochemical properties, which is both a reactive and preventive

---

[96] Staniforth, J. COVID-19 update: worker health, absenteeism present largest risks to U.S. food supply chain [EB/OL] https://www.foodqualityandsafety.com/article/covid-19-update-worker-health-and-absenteeism-present-largest-risk-to-u-s-food-supply-chain/ 2020/2020-03-20.

[97] OECD. Managing international migration under COVID-19 [EB/OL]. http://www.oecd.org/coronavirus/policy-responses/managing-international-migration-under-covid-19-6e914d57/ 2021/2020-08-05.

[98] OECD. Supporting livelihoods during the Covid-19 crisis: closing the gaps in safety nets [EB] http://www.oecd.org/coronavirus/en 2020/2020-08-05.

approach."[99] Cold chain system includes precooling/freezing, cold storage of chilled and frozen food, refrigerated logistics transportation. According to resources, "people can become infected by exposure to a package contaminated with Covid-19 live virus"[100] as low temperature environment prolongs the survival of the virus. Hence, at present different tests are conducted to detect the virus on food to ensure the consumption of safe food. The in-depth analysis of food supply chain suggests that Covid-19 has imposed potential risks on the food supply in terms of available and access to sufficient quantity of nutritious and safe food.

According to International Labour Organization (ILO) report (2020), the Covid-19 crisis had a greater impact in the labor market and on income, mostly because of the drop in economic activity. This drop-in economic activity leads to increased food insecurity within and across countries.[101] This indicates that parallel to decrease in income, food price increases with decline in food choice directly affects access to food in terms of affordability. So, the impact of Covid-19 crisis from economic perspective is equally evident. International Monetary Fund (IMF) (2020) projects a 5% decline of the world economy in 2020—a much deeper global recession than during the global financial crisis of 2007–2009. However, compared with the 2007–2009 food price crisis, the food insecurity concern during Covid-19 crisis is mainly driven by supply disruptions and logistics constraints because of the restrictive Covid-19 policy which affected both food supply and demand with uncertain effects on food prices. So to mitigate such disruptions reviewing the existing trade policies, such tariff barriers, transportation systems, and so forth is essential. Recently FAO reported a 20.7% rise in global food prices driven by dairy and vegetable oils. Before the impact of Covid-19 crisis on food security which exacerbated food insecurity is overcame Russian Federation and Ukraine crisis 2022 emerged which had even greater impact on food insecurity as it increased the food price even greater as Ukraine and Russia make up around 80% of sunflower oil exports.[102] According to the recent report jointly released by Five UN agencies the number of people affected by hunger in the world in 2021 has reached 828 million, an increase of about 46 million from 2020, and a cumulative increase of 150 million since the Covid-19 crisis.[103]

Analyzing the three crisis times and its impact on food security it can be asserted that any nature of crisis has a greater, internal and external, impact on the global food supply chain which eventually increases the food prices disrupting food affordability,

---

[99] Zhu, X., Yuan, X., Zhang, Y., Liu, H., Wang, J., & Sun, B. The global concern of food security during the COVID-19 pandemic: Impacts and perspectives on food security [J] Food chemistry, 2022 (370): 130.

[100] Ibid.

[101] Smith MD, Wesselbaum D. COVID-19, Food Insecurity, and Migration [J] J Nutr. 2020, Vol. 150(11):2855–2858.

[102] Skelding, Conor, Food prices reach record highs, skyrocket by 20.7 percent across globe [N] New York Post, 2022-03-05.

[103] FAO, IFAD, UNICEF, WFP and WHO. The state of food security and nutrition in the world 2022 [R] Rome: FAO, 2022.

food accessibility, and ultimately exacerbates food insecurity. Consequently, food insecurity can accelerate global migration.[104]

Hence, the importance of policy framework, cooperation, and food governance to mitigate the effects of any potential crisis, such as climate change, disasters, wartime disasters, on food security cannot be overlooked; as it will require market control, food supply control, and food aid assistance. So, in pursuance of establishing risk management and control system under The Treaty following aspects can be considered while drafting which can be established both at international and national levels:

1. **Risk management goals and objective:** should ordinarily describe in simple terms the technique of controlling and avoiding potential threats or risks to food security. So, the following might be considering as the objective[105]:
   - (I) Identifying, analyzing and evaluating main potential threat or risk to food security, so aim at anticipating main potential threat or risk.
   - (II) Identifying the main events and situations that could have a significant impact on food security, set measures for controlling and mitigating risks, and facilitating the attainment of the objectives of risk management system and the Law of Food Security.
   - (III) Guide for the implementation of the risk management measures and verify.
   - (IV) Monitoring and reviewing risk management measures and outcomes for continuous improvement by learning from past experiences.
   - (V) Supports sufficient use of resources, assures stakeholders, continuity, transfer of technology and better communication.
2. **Establishing Organizational framework:** a framework that defines roles and responsibilities sets out the procedures and clear and consistent standards for the system. Such as, establishing executive management or the executive board, audit committee, risk manager, and so forth.
3. **Establishing four-stage risk management functioning system:** the three stage functioning system should consider:
   - (I) **Risk identification:** This involves centralizing the main potential risks in the attainment of the objective of The Treaty analyzing an event, one or more sources, and one or more consequences.
   - (II) **Risk analysis and assessment:** This stage involves examining and assessing extends or severity of the potential food security risks both in terms of availability of food and consumption of safe food.

---

[104] See, Sadiddin, A., Cattaneo, A., Cirillo, M. et al. Food insecurity as a determinant of international migration: evidence from Sub-Saharan Africa [J] Food Sec. 2019 (11): 515–530; See also, Orjuela-Grimm M, Deschak C, Aragon Gama CA, Bhatt Carreño S, Hoyos L, Mundo V, Bojorquez I, Carpio K, Quero Y, Xicotencatl A, Infante C. Migrants on the Move and Food (In)security: A Call for Research [J] J Immigr Minor Health 2022,24(5):1318–1327.

[105] For drafting the goals and objectives of the risk management and control system. Food safety Risk Management evidence-informed policies and decisions, considering multiple factors, FAO guidance materials are considered.

(III) **Risk management procedures:** In this stage, the most appropriate action plan or plans involving several measures to maintain acceptable risk levels, i.e. reducing, or eliminating risk, and considering the potential effects and consequences of the risk develop mitigating measures.

(IV) **Monitoring and reviewing risk management measures and outcomes:** This stage helps to comprehend success of the risk management action plan or plans.

4. **Operational procedures and methods:** This specify how an action or process should be carried out within a given time-frame, such as, operating/reporting lines, decision-making and assessment tools, control frequency, person responsible for control, and so forth.
5. **Establish Food Aid Assistance:** Every Contacting Parties under risk management and control system should establish food aid policies from two perspectives: as donor State and as recipient State. According to FAO Right to Food Guideline, "Donor States should ensure that their food aid policies support national efforts by recipient States to achieve food security, and base their food aid provisions on sound needs assessment, targeting especially food insecure and vulnerable groups. In this context, donor States should provide assistance in a manner that takes into account food safety, the importance of not disrupting local food production and the nutritional and dietary needs and cultures of recipient populations. Food aid should be provided with a clear exit strategy and avoid the creation of dependency. Donors should promote increased use of local and regional commercial markets to meet food needs in famine-prone countries and reduce dependence on food aid."[106] Furthermore, bilateral food aid should be consistent with the FAO Principles of Surplus Disposal and Consultative Obligations, the Food Aid Convention and the WTO Agreement on Agriculture, and should meet the internationally agreed food safety standards, bearing in mind local circumstances, dietary traditions and cultures.

6.3.10 Prohibition of Starvation as a Method of Warfare

Prohibition of starvation as a method of warfare is a critical step toward addressing one of the most brutal and inhumane tactics of modern conflict: the deliberate use of starvation as a weapon of war. While international humanitarian law (IHL) already prohibits attacks on civilians, this provision must uniquely and explicitly codify starvation as a distinct war crime, strengthening accountability and closing legal gaps. Such as defining it as: "Starvation of the civilian population as a method of warfare is absolutely prohibited. This prohibition extends to any act or omission that intentionally and foreseeably deprives civilians of objects indispensable to their survival, including, but not limited to, foodstuffs, water, medical supplies, and shelter." This

[106] FAO. Voluntary Guidelines to support the progressive realization of the right to adequate food in the context of national food security. [R] Rome: FAO, 2005.

will move beyond general prohibitions to provide a clear, detailed framework for identifying and prosecuting such atrocities.

Secondly, it's important to define the prohibition acts which should not be limited but must go beyond mere food destruction to encompass the wilful impeding of humanitarian aid, a tactic frequently used to starve populations without direct violence. By including the use of sieges and blockades, it directly addresses a common strategy that has devastating consequences in conflicts like those in Gaza, Yemen, and Sudan. Thirdly, it is also important to extend individual criminal responsibility to military commanders and other superiors, ensuring that those who give orders or fail to prevent these crimes are held accountable. This moves the focus from a state-level violation to individual culpability, which is a cornerstone of modern international justice.

For this provision to be fully developed and effective, several key steps need to be taken. It would require broad adoption by the international community, likely through a new international treaty or an amendment to the Rome Statute of the International Criminal Court (ICC). The inclusion of specific consequences, such as mandatory international sanctions and reparations, adds teeth to the prohibition, providing a crucial deterrent that goes beyond simple condemnation.

6.3.11 *Establishing Food Knowledge Innovation and Management System at National and International Levels*

Knowledge innovation, process, management, sharing, communication, and application are key factors in improving the food insecurity situation. Knowledge innovation in building food security is one of the major key factors as it by utilizing experimental research and development activities and empirical practice activities it promotes generation of new knowledge that might include technical innovation and system innovation which contributes in the food supply chain. For instance, to implement the "double carbon – dual policy" principle recommended under the Treaty knowledge innovation in terms of technical innovation and system innovation is essential. Similarly, to increase food production, nutrition value of food, and reducing food waste through the food supply chain knowledge innovation is indispensable. So, continuous knowledge innovation is indispensable for the global food system to build and maintain sustainability and operate a green food system. Knowledge innovation leads to knowledge process which involves four (4) stages: knowledge generation or innovation, acquisition, construction, and codification.[107] This knowledge process

[107] See, Huber, G.P. Organizational learning: The contributing processes and the literatures [J] Organization Science, 1991, Vol. 2(1):88–115; See also, Pentland, B.T. Information systems and organizational learning: The social epistemology of organizational knowledge systems [J] Accounting, Management& Information Technology, 1995, Vol. 5(1): 1–21; See also, Grover, V. & Davenport, T.H. General Perspectives on knowledge management: Fostering a research agenda [J] Journal

is a part of knowledge management which involves three (3) phrases: (1) four (4) knowledge process; (2) storage, distributions, and transferring; and (3) application. So, the first phrase begins with knowledge generation or innovation which "includes all processes involved in the acquisition and development of knowledge"[108]; then, knowledge acquisition which refers "to the processes through which organizations obtain knowledge externally and create knowledge internally"[109]; later, knowledge construction which refers "to the process of adding or replacing new material within the collective stock of knowledge."[110] Lastly, stage of the first phrase is knowledge codification which "involves the conversion of knowledge into accessible and applicable formats."[111] The second phrases three (3) main phrases, beginning with knowledge storage/retrieval which refers to knowledge being stored in the organizational memory and retrieved to influence activities; then, knowledge distributions which involves "the process by which knowledge is disseminated to places where it is needed and it can be applied"[112]; last stage of the second phrase is knowledge transferring which "includes the movement of knowledge from its point of generation or codified form to the point of use and among individuals and groups."[113] The third phrase involves knowledge application which involves the process "knowledge being applied to create organizational capabilities to achieve performance improvements."[114] To establish and maintain a sustainable and green food system, the Treaty recommends the establishment of a knowledge innovation and management system at the national and international levels. This will facilitate cooperation, monitoring, and review of the global food insecurity situation. Establishing a Food Knowledge Innovation and Management System at both national and international levels may involve the following considerations:

---

of Management Information Systems, 2001, 18(l): 5–21; See also, I. Nonaka A dynamic theory of organizational knowledge creation [J] Organization Science, 1994, Vol. 5(1): pp. 14–37; See also, Alavi, M. & Leidner, D.E. Knowledge management and knowledge management systems: Conceptual foundations and research issues [J] MIS Quarterly, 2001, 25(l): 107–136.

[108] I. Nonaka A dynamic theory of organizational knowledge creation [J] Organization Science, 1994, 5(1) (1994): pp. 14–37.

[109] Huber, G.P. Organizational learning: The contributing processes and the literatures [J] Organization Science, 1991, Vol. 2(l):88–115.

[110] Pentland,.B.T. Information systems and organizational learning: The social epistemology of organizational knowledge systems [J] Accounting, Management& Information Technology, 1995, 5(1): 1–21.

[111] I. Nonaka A dynamic theory of organizational knowledge creation [J] Organization Science, 1994, 5(1): pp. 14–37.

[112] Pentland, B.T. Information systems and organizational learning: The social epistemology of organizational knowledge systems [J] Accounting, Management& Information Technology, 1995,5(1): 1–21.

[113] Alavi, M. & Leidner, D.E. Knowledge management and knowledge management systems: Conceptual foundations and research issues [J] MIS Quarterly, 2001, 25(l): 107–136.

[114] Ibid.

1. **Developing a framework for knowledge management:** This should include defining roles and responsibilities, setting up procedures, and establishing standards for managing knowledge related to food security.
2. **Identifying knowledge gaps and opportunities:** This involves identifying areas where there is a lack of knowledge or where new knowledge can be developed to enhance food security.
3. **Developing strategies to acquire and share knowledge:** This includes developing a system for collecting, storing, and sharing knowledge related to food security, such as through research, innovation, and capacity building.
4. **Encouraging collaboration and partnerships:** Collaboration and partnerships among stakeholders, including governments, academia, research institutions, and industry, can facilitate the sharing of knowledge and innovative practices.
5. **Building capacity for knowledge management:** This includes developing training programs to build knowledge management skills and awareness, and supporting research and development to enhance food security.
6. **Establishing monitoring and evaluation mechanisms:** This includes developing indicators to measure the effectiveness of the knowledge management system and the impact of the acquired knowledge on food security.

Because of the continuous knowledge innovation which is indispensable for the global food system to build and maintain sustainability and operate a green food system, The Treaty encourages the State Parties to establish "knowledge innovation and management system" at national level under FSD and at international national level through organizational support; which will build cooperation, and help to monitor and review food insecurity situation at global level.

6.3.12 Dispute Settlement Mechanism

Dispute Settlement Mechanism (DSM) is an essential part of any international treaty or agreement which contributes to resolve any dispute on whether a country has acted in conformity with its international obligations, and further contributes in developing the interpretative understanding of the treaty or agreement. As The Treaty creates a forum to form deeper integration among the State parties, whose benefits can only be realized if the Treaty provisions implemented accordingly and faithfully. So, disputes will inevitably arise regarding the scope, nature, and interpretation of the Treaty and DSM shall contribute in resolving such conflicts of contentious nature.

Conventional such DSM exists in the international area in several forms such as: The International Court of Justice (ICJ), the International Tribunal for the Law of the Sea (ITLOS), WTO Dispute Settlement Mechanism, and so forth. Considering the amount of pending casing before the existing international courts and tribunals along with the nature particular nature of the conflicts that might arise under the Treaty, the Treaty adhering to ITLOS encourages the establishment of its own dispute settlement mechanism, the International Tribunal for Food Security (ITFS).

As disputes regarding food security are an emerging zone, hence, ITFS requires specialized expertise on the international food law, international law, international legal system, and so forth, to resolve the dispute in a peaceful manner. ITFS as an independent judicial body established under The Treaty shall adjudicate disputes arising out of the interpretation and application of the Treaty and shall be open to State parties to the Treaty, intergovernmental organizations which are not parties to the Treaty, and to state enterprises and private entities. So, ITFS shall have jurisdiction over any dispute concerning the interpretation or application of the Treaty, and over all matters specifically effecting food security and exacerbating food insecurity. Unlike the UNLOS, The Treaty does not refer to any other dispute settlement forum other than ITFS, which is a compulsory dispute settlement entailing binding decisions. However, because of the considerable experience and knowledge ICJ holds, it could be passed to ITFS.

The further procedure of the ITFS can be developed by studying the theory under which the existing international dispute settlement mechanism operates.

6.3.13 Food Policy Reviewing Mechanism

The recommended "food policy reviewing mechanism" (FPRM) is not an emerging or a newly recommend mechanism. WTO holds a similar mechanism named as Trade Policy Review Mechanism ("TPRM") and Annex 3 of the Marrakesh Agreement forms the legal basis for the mechanism. So FPRM is an important component of global efforts to promote global food security. They provide a means of assessing existing food security agreements, implementing sound policies to promote food security, and addressing humanitarian and environmental issues in food security. The main objective of the FPRM should be to contribute to improved adherence by all Contracting Parties of the Treaty to rules, disciplines, and commitments made under the Treaty and its application by carrying out regular collective review, assessment and evaluation of individual Contracting Party's food policy and food security practices for the smoother functioning of the global food security system, to achieve greater transparency in, and understanding of, the food policies and practices of Contracting Parties.[115]

In other words, FPRM will enable countries to evaluate their existing food provisions and to identify areas where improvements can be made to better guarantee food security for all. The review enables countries to compare and contrast their own national food security policies with those of other countries. This comparison can go beyond simple economic variables to analyze how policies and programs affect nutrition, poverty, health, educational attainment, and access to water and sanitation.

The review will also provide an opportunity to evaluate the effectiveness of existing the Treaty provisions. It can help identify strengths and weaknesses and

[115] The objective of the FPRM is derived and inspired from WTO TPRM.

where modifications may be needed in order to meet eventual goals of food security. By examining treaty language, it can help find potential gaps and problems. By looking at national laws and regulations, it can assess member states' commitment to implement the terms of their treaties and make sure that implementation is done effectively.

In addition to assessing existing food policy, the review mechanism will help countries anticipate and address potential humanitarian and environmental problems that could arise from food insecurity. It is important for countries to recognize the potential effects of a food security challenge in order to take appropriate measures to protect vulnerable populations from potential harm. By using a review mechanism, countries can identify ways to reduce the risk of the effects of food insecurity, such as malnutrition and poverty, while maximizing existing resources. The review mechanism will contribute to evaluate the effect of current the Law of Food Security. In particular, it can be used to determine the success of international donations, partnerships among global food systems, and the role played by international organizations.

Considering its objective FPRM requires a Food Policy Review Body (FPRB) to carry out the collective review, assessment, and evaluation of individual Contracting Party's food policy and food security practices. Moreover, the food policies and practices of all Contracting Parties shall be subjected to periodic review considering four factors: (1) in terms of food security, both food availability and food standard, and food waste at national level; (2) its impact on the global food security system; (3) its impact on food market stability; and (4) in terms of the Law of Food Security. So to conduct such as assessment and evaluation of individual Contracting Party's food policy and food security practices, FPRB needs to develop a basic plan, consultation, and report which can be presented in the UNGA.

A review mechanism is necessary to ensure that The Treaty provisions are kept up to date, consistently enforced, and continually improved. The review mechanism will help ensure that all countries are doing their part to promote food security and to protect vulnerable communities from the effects of a food insecurity challenge. It is an important tool for global efforts to reduce hunger and to bring about a fairer and more equitable food system.

6.3.14 Reciprocal and Non-Reciprocal Food Security Agreement

As the Treaty is a multilateral treaty which encourages the State Parties to strength the national food system to bring a collectively enhance the global food system, hence, legal cooperation is indispensible to build a sustainable and equitable global food system. A sustainable and equitable global food system requires knowledge innovation, sharing, and advance green technological support through international technology transfer or diffusion which requires legal win-win cooperation between developed and developing countries. International technology transfer or diffusion and knowledge diffusion are essential for a sustainable and equitable global food system, firstly, it allows one country to gain access to and employs technology developed in another country; and secondly, dissemination of knowledge supports other potential inventors. However, the need of stronger IPR protection in this regard cannot be ignored.

Considering, the importance of building a sustainable and equitable global food system through partnership, the Treaty encourages its State Parties to undertake "reciprocal or non-reciprocal food security agreement" among them, due to its the diverse nature, one common feature, and legal cooperation. Reciprocal food security agreements are those in which two or more countries agree to provide each other with a mutually beneficial arrangement. This could involve any number of arrangements, such as the exchange of goods, the mutual support of farming practices, the mutual sharing of technology and agricultural expertise, or the provision of funds from one country to another. These types of agreements are usually negotiated by the governments of both countries, with the intention of creating a stable and durable food security system between them.

Non-reciprocal food security agreements are those in which one party provides benefits to another without any expectation of any form of return. This is often done through the donation of food goods, the provision of funds for agricultural and technological development, or the provision of agricultural expertise. This type of agreement is mainly implemented by developed countries to help those in the developing world establish a stable food system.

Both reciprocal and non-reciprocal food security agreements are highly effective at ensuring the stability and availability of food across different countries. In addition, they provide a platform for countries to mutually benefit one another and strengthen their relationship in times of crisis. However, there are also some potential drawbacks. Namely, food security agreements can lead to a lack of competition and ignorance as to the true underlying costs and benefits of food production and distribution. If these agreements are not properly monitored, it can lead to high prices, low quality goods, and a decreased incentive to innovate.

Latest research suggests the impact of trade agreements on trade flows from developed countries to developing countries, or vice-versa, reciprocal and non-reciprocal agreements have had a positive effect.[116] So, "reciprocal or non-reciprocal food security agreement" with diverse nature and one common aim of developing a sustainable and equitable food system and eradicate food insecurity can form a legal cooperation. Furthermore, The Treaty encourages "reciprocal or non-reciprocal food security agreement" to be conducted specially between developed (provider) and developing countries (beneficial) for a certain period of time which is reasonable to improve food security situation by building resembling national food system through investment, international technology transfer, or diffusion and knowledge diffusion. For instance, China Africa Cooperation—"China in Africa Project" (the Project) (2021). One of the aims of China and the African Union is to boost bilateral cooperation

[116] Gil-Pareja S, Llorca-Vivero R, Martínez-Serrano JA. Reciprocal vs nonreciprocal trade Agreements: Which have been best to promote exports? [J] PLoS ONE, 2019, 14(2).

on food safety and security. So, China partners with the continent to ensure Africa achieves zero hunger-one goals of the UN 2030 Agenda for Sustainable Development.[117] Under the project China has undertaken 142 agricultural projects which are gradually showing fruitful results in developing the national food system in Africa. Anecdotal evidence shows the Project could contribute to agricultural growth and poverty reduction in Africa. As China's own successful experience in agricultural growth and poverty reduction could provide valuable lessons for Africa.[118] Hence, The Treaty encourages its State Parties to collaborate jointly within the framework of this Treaty with the aim to strengthen the global food system.

Reciprocal and non-reciprocal food security agreements are a necessary step in ensuring a stable food supply for both the immediate and long-term future. However, it is important that these agreements are carefully monitored and regulated, in order to prevent any potential drawbacks from taking place. Doing so, countries are better able to guarantee that all individuals, communities, and nations have access to adequate, safe, and nutritious food at all times.

6.4 Challenges of the Law of Food Security

In the construction of global food security mechanism, from the perspective of the United Nations Millennium Development Goals and 2030 SDGs, looking forward to the solution of food insecurity problem. And put forward the more specific principles and rules involved in the proposed Law of Food Security.

The Treaty is designed to ensure access to safe and nutritionally adequate food for all people. It sets the standards for food availability, accessibility, and quality to protect people's right to adequate food and nutrition. This right is recognized in international human rights law, but is often overlooked domestically. Hence, enforcement of The Treaty might be a major challenge faced by governments and international organizations. The enforcement of The Treaty requires a two-pronged approach: one to ensure compliance by states and international organizations, and another to ensure compliance by non-state actors such as corporations and non-governmental organizations. This makes enforcement of the Treaty an inherently complex process.

At the state level, enforcement of the Treaty primarily relies on cooperation between countries. States must work together to monitor and enforce compliance with the treaty. This can be a challenge due to different levels of commitment, capacity, and willingness to act. Furthermore, countries also have different food security standards and definitions. As such, this can make it difficult to enforce the Treaty across a wide variety of countries.

---

[117] Munene, Dennis. Sino-African teamwork improving food security [N] China Daily 2019-12-09.

[118] Lei Sun, Helen. Understanding China's Agricultural Investments in Africa. South African Institute of International Affair [EB/OL] www.files.ethz.ch/isn/146037/saia_sop_102_lei_sun_20111129.pdf 2011-11-29/2020-05-06.

At the non-state level, enforcement of the Treaty is complex and challenging. Non-state actors such as corporations and NGOs are often outside the reach of international law. As such, these actors are not subject to the same legal requirements as states and can ignore the Law of Food Security. Furthermore, these actors are often operating in countries where there is little or no regulation to ensure their compliance with the treaty. This makes it difficult to ensure they are held accountable for violations.

To ensure effective enforcement of the Treaty, governments and international organizations must take an integrated approach. This includes states working with each other to ensure compliance and promote better food security, and commercial actors being held to the same standards as governments. Additionally, strengthening the capacity of international actors to monitor and enforce the Treaty is essential. This includes providing technical and financial support to states and non-state actors, and engaging in dialogue and capacity building to improve understanding and enforcement of the treaty. To ensure effective enforcement, governments and international organizations must take an integrated approach that includes states working with each other and the holding of non-state actors to the same standards. This must be accompanied by strengthened capacity to monitor and enforce the Law of Food Security, with the provision of technical and financial support.

6.5 Conclusion

Food security is a global concern due to its implications on human health and well-being. It is particularly important to ensure that people everywhere have access to sufficient, safe, and nutritious food as well as the means to consume it. To ensure food security and ensure that people everywhere have access to it, countries have to sign certain agreements and treaties, such as the Law of Food Security.

The Treaty as an international multilateral treaty among developing and developed countries will play a vital role as a leg commitment to guaranteeing access to food for everyone. This Treaty is based on some major fundamental principles which put forth the goals of protecting human rights to food, promoting cooperative and sustainable food production, and directing international food aid at the most vulnerable and food-insecure populations. The treaty also contains a number of specific commitments, such as the commitment to fighting global hunger.

The Treaty requires countries to implement policies that promote sustainable agricultural production and discourage further land degradation. To do this, signatory states must commit to minimum standards for environmental responsibility, including the use of sustainable practices such as crop rotation, integrated pest management, and soil conservation. The Treaty also establishes a framework for food trade regulations and strengthens efforts to ensure the availability and affordability of food products.

One of the primary goals of the treaty is to reduce hunger and ensure everyone has access to a nutritious diet. The Treaty requires countries to reduce their reliance on food imports and develop self-sufficiency in food production. Furthermore, states must tackle food access issues by creating comprehensive food policies, such as price controls, subsidies, and food aid measures. In addition, the treaty encourages governments to make greater investments in nutrition education and to support the efforts of local food producers and vendors.

In addition to the stated regulations, the Treaty also outlines mechanisms for monitoring and evaluating countries' compliance. Countries are encouraged to use global satellite imagery data to track crop harvest successes and food flows. This data can then be used to track vulnerable populations and highlight areas where support is needed. In addition, signatory countries must follow-up on any reported human rights violations relating to food access and production.

The importance of the Treaty cannot be overstated. By committing to these principles, states are able to take tangible action toward the elimination of hunger and the establishment of food security. The treaty further recognizes that food security is a fundamental human right and that everyone, regardless of occupation or social class, is entitled to an adequate, safe, and nutritious diet. When implemented, the Treaty can help to improve access to food.

The Treaty also sets a solid foundation for the international community to work together in fighting global hunger. It sets global standards for food security and emphasizes the importance of cooperation and collective action in the promotion of food security. In doing so, it will create an opportunity for countries to collaborate, learn from each other, and share resources in the fight against hunger.

The Treaty will be an important step toward providing access to food for people everywhere. It will help to set global standards and hold states accountable for their respective commitments. Most importantly, it assures that all people everywhere have the right to access to safe, affordable and nutritious food. As such, it is a critical tool in strengthening food security and achieving the goal of a world with zero hunger. UN Secretary General António Guterres in October 2019 acknowledging the dire conditions of the contemporary global food system stated, "Today there is widespread consensus among a growing number of scientists, civil society organizations, and governments that the global food system cannot be sustained in its current configuration — economically, ecologically, or socially."[119] In order to build a sustainable and equitable global food system, this chapter provides an outline for the Law of Food Security; developed with the view to guide international organizations, policymakers, and national governments to build a sustainable and equitable global food system. The Treaty requires implementation at both national and international levels. So, being a multilateral treaty, the Treaty needs to be open for signature, ratification, or accession under the umbrella of UN, its relevant Agencies, intergovernmental organizations, and other relevant organizations such as FAO, WHO, and WTO.

[119] Montenegro de, et al. Resetting Power in Global Food Governance: The UN Food Systems Summit. [J] Development, 2021, 64(3–4): 153–161.

References

Alinovi L, Hemrich G, Russo L (2008) Beyond relief: food secuirty in protracted crises. New Internationalist

ILO (2020) COVID-19 and the world of work: Updated estimates and analysis 3rd ed. Switzerland. https://www.ilo.org/wcmsp5/groups/public/@dgreports/@dcomm/documents/briefingnote/wcms_743146.pdf. Accessed 10 Aug 2020

IMF (2020) World economic outlook update: A crisis like no other, an uncertain recovery https://tinyurl.com/yazvtk7d. Accessed 10 Aug 2020

Kirk, Jeffrey W (1994) General principles of Law. Annual Survey of International & Comparative Law, Vol. 1(1)

McGee H (1997) On food and cooking. New York: Scribner

Renn O, Dreyer M (eds) (2009) Food safety governance. Springer, Berlin Heidelberg. https://doi.org/10.1007/978-3-540-69309-3

Shelton D (2017) Common concern of humanity. In: Globalization and common responsibilities of states. Routledge, pp 3–10. https://doi.org/10.4324/9781315254135-1

Vincent B, Bado, André Bationo (2018) Integrated management of soil fertility and land resources in Sub-Saharan Africa: involving local communities. In Donald L. Sparks (eds.) Advances in Agronomy Volume 150:1–33. https://doi.org/10.1016/bs.agron

WFS (1996) World Food Summit 1996, Rome declaration on World Food Security. World Food Security

World Bank (2018) Arable land (% of land area) . Data. data.worldbank.org/indicator/AG.LND.ARBL.ZS/. Accessed 20 Feb 2020

Conclusion and Outlook

The food system we inherit is designed around the middle of the twentieth century when the question of food insecurity was discussed in the simplest terms considering it as a national issue. However, population growth and insufficient productivity growth were continuously challenging the food system (Schutter, 2014). The core argument of this thesis is that the global food system in its current configuration is not economically, ecologically, or socially sustainable and equitable. Mostly, because the contemporary international instruments regulating the contemporary global food system focuses on food safety standards, not focusing on food security in terms of sufficient quantity of food, and operates an unfair global food system with diverse and conflicting food safety standards which presents the disparity between developed and developing countries. This thesis has discussed in-depth the pressing need of building a sustainable and equitable global food system. With that aim, Chap. 1 presents the theoretical framework to connect the research to existing knowledge and provide an explicit statement of theoretical assumptions by proposing well-substantiated explanations. Presenting the status of food security, Chap. 1 emphasized the fact that scholars engaging in the literature have invariably identified five issues challenging the global food system: (1) disparity in comprehending the notion FS and FIS; (2) interrelation between trade policy; food security and international food law; (3) food waste and food security; (4) interrelation between food security and climate change; and (5) ensuring food rights, safety and governance in the global food system. The scholars have analyzed contemporary international food law focusing on food safety standards in food trade and do not address other food security issues. So, the scholars concentrated primarily on understanding the food security under economic perspective and understanding the disputes from food safety perspective.

Chapter 1 identified eight (8) major problems: (1) The approach toward food security; (2) Lack of Food Governance. (3) Unequal distribution of food; (4) Lack of Corporation and Harmonization; (5) Absence of International Food Law; (6) No anti-food waste provision; (7) Absence of application of legal principles in the global food system; and (8) Lack of Cooperation, Collaboration and Harmonization.

S. S. Silvee, *The Law of Food Security: The Legal Framework for a Sustainable and Equitable Global Food System*, LITES - Legal Issues in Transdisciplinary Environmental Studies 5, https://doi.org/10.1007/978-981-95-3636-8

Chapter 2 highlighted the interrelation between FIS, FS, and safety which contributed in the analysis of the contemporary international food law in Chap. 5 to identify targeted issues and probable effective ways for food policy interventions.

Chapter 3 examines one of the major research questions whether or not food wastes is exacerbating food insecurity? If yes, is there anti-food waste system within the framework of the major contemporary international food law governing the global food system? The findings of the chapter asserts that the average food waste across the global food supply chain per year is around 36 percent and sum-up of the average estimated percent of food waste in per year is around 51 percent. The results of the "statistical analysis" denoted that out of 94.6 percent of people visiting restaurants $\geq$74.4 percent wastes food of an amount ($\approx$) 100 g food at restaurants in each visit, the quantity of food wasted can feed approximately twenty-six (26) people once a day. The study also estimated that $\geq$22.51 percent of people visiting restaurant wastes food in each visit. The findings of the chapter substantiated that food waste occurs though the food supply chain, from production to consumption, exacerbating food insecurity. The chapter showed there is an association between food waste and exacerbating food insecurity. Hence, anti-food waste system is essential in the global food system. Furthermore, for measuring what percentage of food waste will contribute in food insecurity and to control the level of food waste, the study developed Food waste severity measuring scale; which can be used by countries and international organizations to measure and control the level of food waste both nationally and internationally. Moreover, Chap. 3 delves into the relationship between food waste and food security, highlighting the need to build a sustainable and equitable food system. Through meta-analysis and statistical analysis of previous studies and web-based surveys, the chapter provides valuable insights into the impact of food waste on the global food supply, attitudes toward food waste across different cultures and regions, and the importance of addressing this issue in achieving food security.

Building on this understanding, Chap. 4 examines the evolution of international food law from the League of Nations to present day. By tracing the development of these instruments, the chapter sheds light on the historical context of food insecurity and the international response toward addressing it, from a shift toward food security post-World War II to the present-day focus on food safety, standards, and waste. Understanding this historical perspective is crucial in comprehending the six components of the global food system and the challenges that persist in achieving food security. It highlighted the how food security and food security has been comprehended from economic and trade perspective. It further demonstrated the shift and changing approach in building food security, from focusing on food insecurity to hunger, malnutrition, and poverty. This change in approach has transformed the notion "food security" in to a complex notion. Further, the chapter highlighted the contemporary international food instruments are drafted based on three major notions food insecurity, food security, and food safety, however, mostly focused on food safety, facilitating food trade, harmonizing food standards and ensuring right to safe food. The chapter emphasized on the fact that the contemporary global food system is depended on scattered binding and non-binding instruments.

Chapter 5 analyzing the three major contemporary international food law, the chapter affirms the findings of historical analysis of international food law i.e. the instruments focus particularly on food safety. It further suggests that the three major contemporary international food law fall short to build a legal mechanism and institutional framework required for operating the global food system to build food security. The chapter emphasized on enacting a multilateral treaty, i.e. the Law of Food Security, to build a sustainable and equitable global food system. Highlighting the need to enact the Law of Food Security, the chapter attempts to provide a basic structure for the multilateral treaty. The study developed a model for global food system in terms of its nature and functionality to offer an institutional framework.

In the light of the arguments and findings presented in the thesis, Chap. 6, proposed an outline of a multilateral treaty, entitled as the Treaty which will not only provide a legal mechanism but will also build an institutional framework that is required to build a sustainable and equitable global food system. The outline is developed analyzing and considering the international and national food, agricultural, food and nutritional policy and legislation of twenty (20) developing and developed countries. The proposed Treaty needs implementation at national and international levels through signature, ratification, or accession under the umbrella of UN and its Agencies, as the treaty promotes collaboration among states in order to efficiently and effectively implements sustainable agricultural practices. This includes sharing of scientific research, technology, and information. Additionally, the treaty encourages global dialogue on food security by creating an international governing body to assess progress.

Overall, the Treaty is an important step toward a more equitable global food system. By implementing the outlined policies and regulations, signatory countries can ensure the availability and affordability of nutritious foods while also implementing sustainable production practices that reduce land degradation. In addition, the establishment of an international monitoring body will help to ensure that countries are following the agreed-upon guidelines and working toward peace-building through food security.

After the epidemic situation, the study of international law on the topic of food security and international food law has important theoretical significance and practical guidance. Ontologically, this study makes a systematic and in-depth comparative study of international food security agreements or instruments of different nature, organization, and effectiveness. The study laid a theoretical foundation for the construction of an effective international Treaty and foreshadows a practical diplomatic road map. From the political or diplomatic standpoint of developed and developing countries, it reveals the challenges faced by the contemporary global food system, in term, of food security and food safety, and analyzes the deficiencies of the existing international legal mechanism. Starting from theoretical coordinates and legal issues, the study focused on the shortcomings of the existing legal mechanism and its theoretical interpretation from diachronic and synchronic perspectives. On this basis, the basic content and internal logic of the Treaty is conceived; especially the sustainability and equity values pursued by its legal system are emphasized. In addition, on the basis of objective statistical analysis, the study discussed the important impact of food waste on food security, and puts forward relevant suggestions.

In conducting the research, this study witnessed some caveats, one of which was access to critical academic resources, developing literature on food security, data deficiency on food waste, and the fact that academic research in this field is still evolving. Another probably limitation is that, the research has conducted survey from consumer's perspective visiting at restaurants, although, survey from restaurants operators' perspective, and other phrases of the food supply chain would have given a more in-depth approach to the study. However, the data of food waste through the food supply chain is drawn from FAO's database on food loss and waste. Probably, anticipated limitation will be regarding the web-based survey of the consumers that concerns the generality and validity of their results, as web-based survey is not suitable to ask open-ended questions hence, the questionnaire needs to be designed in such a way that it is not lengthy and time consuming.

Despite these caveats, the study makes use of its advantages to carry out theoretical argumentation, uses its shortcomings to discover important legal issues in this field, and puts forward innovative viewpoints and uses corresponding research methods. The study provides useful insights essential for policymakers and different stakeholder involved in the global food supply chain to formulate and implement the Law of Food Security. Despite introducing some challenges, the methodology proposed creates "evidence-based research" portraying how successful the present international food law are governing the modern global food system. Moreover, the research also creates area for further research which can be developed based on the findings of this research to address unanswered aspects. Such as, the Treaty needs further analysis regarding its implication and applicability, so needs recommendations and suggestions. In this regard further research from social-legal aspect can be conducted to comprehend the implication of the Law of Food Security. Additionally, this study addressed that there is insufficient amount of data measuring food waste in food services level so further research can be conducted in this field. This study has a scope for further policy research from multidisciplinary aspect such as food security and climate change, food crime, knowledge management in global food system, food dispute settlement, and so forth.

References

Schutter OD (2014) The right to adequate nutrition. Development 57(2):147–154. https://doi.org/10.1057/dev.2014.64

Appendices

Appendix A: List of International Documents[1]

FAO	United Nations	World Food Council
(1979) Plan of Action on World Food Security. FAO Council Resolution 1/75	(1976) Development and International Economic Co-operation. Resolution 3362 (S-VII) adopted by the UN General Assembly During its Seventh Special Session, 1–16 September. UN General Assembly Official Records, Supplement No. 1 (A/10301)	(1975) Programme of Work on the Council. Note by the Executive Director. Document WFC/4
(1980) Objective Indicators to Signal Acute and Large-Scale Food Shortages. Document CFS: WP/80/4	(1980) International Development Strategy for the Third United Nations Development Decade. UN General Assembly resolution 35/56 adopted on 3 December 1980	(1975) Report of the World Food Council on its First Session. Official Record of the General Assembly, Thirteenth Session. Supplement No. 19 (A/10019) (New York: United Nations)
(1980) Ways of Improving National Preparedness to meet Acute and Large-Scale Food Shortages, Including Improvements in Internal Distribution Networks. Document CFS: WP/80/5	(1987) Environmental Perspectives to the Year 2000 and Beyond. UN General Assembly Resolution 42/186 adopted on 11 December	(1976) International System of Food Security. Report by the Executive Director. Document WFC/22

(continued)

[1] This list of international documents is attached in reference to Sect. 4.2.3 of Chap. 4.

S. S. Silvee, *The Law of Food Security: The Legal Framework for a Sustainable and Equitable Global Food System*, LITES - Legal Issues in Transdisciplinary Environmental Studies 5, https://doi.org/10.1007/978-981-95-3636-8

(continued)

FAO	United Nations	World Food Council
(1981) Requirements of Developing Countries for Food Security Assistance: Storage Needs for Cereals. Document CFS:81/5	(1987) Co-ordination in the United Nations and the United Nations System. Report by the United Nations Secretary-General. Document A/42/232	(1976) Report of the World Food Council on its Second Session. Official Record of the General Assembly, Thirty-First Session. Supplement No. 19 (A/31/19) (New York: UN)
(1982) Regional and Sub-Regional Food Security Schemes aimed at Strengthening the Collective Self-Reliance of Developing Countries. Document CFS: 82/ 4	(1989) Operational Activities for Development. Document A/C.2/44/L.87	(1977) Progress Towards Increasing Food Production in Developing Countries. Document WFC/36
(1982) Director General's Report on World Food Security: A Reappraisal of the Concepts and Approaches. Document CFS: 83/4	(1990) International Development Strategy for the Fourth United Nations Development Decade UN General Assembly Resolution 45/199 adopted on 21 December 1990	(1977) Report of the World Food Council on the Work of its Third Session. Official Record of the General Assembly, Thirty-Second-Session. Supplement No. 19 (A/32/19) (New York: United Nations)
(1983) World Food Security Compact. Document CFS/8. Eighth Session, Committee on World Food Security		(1978) Report of the World Food Council on the Work of its Fourth Session. Official Records of the General Assembly, Thirty-Third Session. Supplement No. 19 (A/33/19) (New York: United Nations)
(1984) World Food Security: Selected Issues and Approaches. Document CFS: 84/4		(1979) World Food Security for the 1980s. Report by the Executive Director. Document WFC/1879/5
(1987) Impact on World Food Security of Agricultural law in Industrialized Countries. Document CFS 87/3		(1979) World Food Security for the 1980s. Plan of Action on World Food Security. Document WFC/1979/5/Add.1
		(1979) World Food Security for the 1980s. International Trade. Document WFC/1979/ 5/Add.2
		(1982) National Food Strategies to Eradicate Hunger

(continued)

(continued)

FAO	United Nations	World Food Council
		(1982) World Food Security and Market Stability. A Developing Country-Owned Reserve. Report by the Executive Director. Document WFC/1982/5
		(1987) Consultation on the Impact of Economic Adjustment on People's Food Security and Nutritional Level in Developing Countries. Document WFC/1987/2/Add.1
		(1988) International Food Trade. Document WFC/1980/5
		(1988) Growing Hunger Amidst Food Surpluses. The Potential for Hunger Reduction Through Food-Surplus Based Development. Report by the Secretariat. Document WFC/1988/2
		(1990) Additional and More Effective Measures by Governments to Alleviate Hunger and Poverty. Report by the Executive Director. Document WFC/1990/3
		(1990) Improving Co-ordination of National and International Action Towards A More Concerted Attack on Hunger. Document WFC/1990/5

Appendix B: Survey Questions[2]

Examining Nexus Between Food Waste and Food Insecurity: A Survey on Food Waste at Restaurants

Hello Everyone,

Food, as a biological necessity, is the only source of energy for human beings and after house the place where maximum people consume food is at restaurants. In order to examine whether food waste at restaurants is a usual habit which can contribute in food insecurity, this survey is being conducted. Please fill out the questionnaire truthfully, as there is no right or wrong answer. Your true answers and opinions will contribute in the study.

Thank you very much for filling out the questionnaire, we promise not to reveal your privacy.

Thank you again for your support and cooperation!

Dr. Sadiya S. Silvee
School of Law, Zhengzhou University.

Part 1 Instructions: Please write down your real information for each.

1. Your nationality: [填空题]*

   ____

2. Your gender: [单选题]*

○ Male	○ Female

3. Your Education [单选题]*

○ Undergraduate
○ Postgraduate

4. Your age range: [单选题]*

○ 18 ~ 25	○ 26 ~ 30	○ 31 ~ 40	○ 41 ~ 50	○ 51 ~ 60	○ above 60

5. Your occupation: [单选题]*

○ Student
○ Work (not a student now)

6. What's your monthly food expenditure? (currency: RMB) [单选题]*

[2] This questionnaire is attached for reference to Sect. 3.6.2 of Chap. 3.

○ Below 800
○ 800–1200
○ 1200–1500
○ 1500–2000
○ Above 2000

Part 2: Food at Restaurants

7. What is type of food you usually prefer? [多选题]*

☐ Vegetables
☐ Fruits
☐ Pork
☐ Other Meat
☐ Fast Food (e.g. burger, pizza, etc.)

8. Do you often go to restaurants or order takeaways? [单选题]*

○ Often
○ Generally
○ Never

9. How often do you eat at restaurants? [单选题]*

○ Less than once a week
○ Once a week
○ 2–4 times a week
○ 5–7 times a week
○ More than 7 times

10. Do you have food waste or leftover? [单选题]*

○ Never
○ Always
○ Often
○ Occasionally
○ Seldom

11. Which meal do you usually have leftovers?[单选题]*

☐ Breakfast
☐ Lunch
☐ Dinner
☐ No leftover

12. Why do you produce leftovers or waste food? [单选题]*

○ Food is not tasty
○ The food is not clean
○ Food is served more/can't eat more
○ Order a lot of food
○ Personal habit
○ Other

13. Have you seen food was wasted in the restaurant by others? [单选题]*

○ Never
○ Always
○ Occasionally
○ Often

14. Do you choose to takeaway your leftover food after eating in the restaurant? [单选题]*

○ Never
○ Always
○ Hesitate
○ Once in a while
○ Often

15. Have you seen others to pack their leftover food after eating in the restaurant? [单选题]*

○ Never
○ Always
○ People hesitate
○ Often

16. What do you think is the main reason why people hesitate to pack their leftover food? [单选题]*

○ Packing leftover food is shameful
○ Worrying about hygiene is problematic
○ Packaged food is not easy to carry
○ Other

17. Which of the following measures do you think can change the status quo of restaurant food waste? [单选题]*

○ Restaurants can offer customers to order meals divided into large and small portions, and the price will change accordingly
○ Restaurants can offer takeaway of leftover food
○ Creating awareness about food waste and letting people know that food is hard to come
○ Fine

Part 3: Food Waste as Global Issue

18. Are you concerned about food waste in general? [单选题]*

○ Yes
○ No

19. Do you think food waste is serious? [单选题]*

○ Very serious
○ Not very serious problem
○ Food waste exists, but it is common

20. Which of the following effects do you think food waste will bring? [单选题]*

○ Waste of resources
○ Environmental pollution
○ Contributes in food insecurity
○ No effects
○ Other

21. Please choose according to your ideas [矩阵单选题]*

	Strongly disagree	Disagree	Neutral	Agree	Couldn't agree more
Food waste will affect the sustainable development of the country and the world adversely	○	○	○	○	○
Food waste will not endanger national food security	○	○	○	○	○
People's food waste will have an adverse effect on social atmosphere	○	○	○	○	○

22. Do you think wasting food is acceptable? [单选题]*

○ Yes
○ No
○ Neutral

23. Do you know that the amount of food produced now is enough for all the people to eat but there're still people in hunger? [单选题]*

○ Yes
○ No
○ Heard of it, but did not care

24. Please choose according to your ideas [矩阵单选题]*

	Strongly disagree	Disagree	Neutral	Agree	Couldn't agree more
I often see advertisements about saving food	○	○	○	○	○
I have seen restaurants and other places to advertise not to waste food	○	○	○	○	○

(continued)

(continued)

	Strongly disagree	Disagree	Neutral	Agree	Couldn't agree more
Schools, workplaces, stadiums, and public gathering or tourist places advertise about not to waste food	○	○	○	○	○
I see social awareness created in social media about food waste	○	○	○	○	○

25. Please choose according to your ideas [矩阵单选题]*

	Strongly disagree	Disagree	Neutral	Agree	Couldn't agree more
I think wasting food is a waste of money	○	○	○	○	○
When I have financial difficulties, I tend to save food	○	○	○	○	○
I rarely think about money when wasting food	○	○	○	○	○
I tend to save more on expensive food than when prices are low	○	○	○	○	○

26. How much food do you think you waste every meal? [单选题]*

○ Unaware/Nothing
○ ≤100 g
○ ≥100 g

27. Food waste while eating with friends or someone [矩阵单选题]*

	Strongly agree	Agree	Neutral	Disagree	Strongly disagree
I often have meals with my friends or with someone	○	○	○	○	○

(continued)

(continued)

	Strongly agree	Agree	Neutral	Disagree	Strongly disagree
Do you think comparing to eat alone, we waste less food when we eat together	○	○	○	○	○

28. Do you think it is necessary to reduce food waste in your life? [单选题]*

○Yes
○No

29. Please choose according to your ideas [矩阵单选题]*

	Strongly disagree	Disagree	Neutral	Agree	Couldn't agree more
I think I have a responsibility to save food	○	○	○	○	○
When I waste food, I feel guilty	○	○	○	○	○
I feel satisfied when I am not wasting food	○	○	○	○	○

30. Are you willing to make some changes in order to avoid food wasting? [单选题]*

○ Yes
○ No
○ It depends__________________

Part 4: Anti-food Waste Law

31. Are you aware or ever heard about any Anti-food waste law? [单选题]*

○ Yes
○ No

32. What do you understand by Anti-food waste Law? [矩阵单选题]*

	Strongly disagree	Disagree	Neutral	Agree	Couldn't agree more
Law to prevent food waste	○	○	○	○	○
Designed to help establish a long-term mechanism to prevent food waste	○	○	○	○	○
Aimed at safeguarding the country's food security	○	○	○	○	○
Specifies the responsibilities of various stakeholders	○	○	○	○	○

33. Which country was the first to ban food waste in stores? [单选题]*

○ Spain
○ France
○ China
○ Unaware

34. Which countries holds Anti-food waste Law? [多选题]*

○ China
○ Spain
○ France
○ Italy
○ Unaware

35. Do you agree adequate government policies can make a huge difference to reduce food waste? [单选题]*

○ Agree
○ Disagree
○ Neutral
○ It depends__________________

36. Are you willing to adhere to Anti-food waste law? [单选题]*

☐ Yes
☐ No
☐ It depends__________________

Appendix C: Twenty Countries and Their Laws on Food and Nutrition[3]

No.	Country name	Legislation and policy
1.	Afghanistan	Afghanistan Food Security and Nutrition Plan 2019 to 2023 Date of text: 12 November 2018 National Horticultural Policy Date of text: 07 March 2018 Afghanistan Food Security and Nutrition Public Awareness and Advocacy Framework and Plan 2018–2023 Date of text: 2018 Food Safety Act Date of text: 19 July 2016 Regulation on Laboratory Analysis of Products of Plant Origin Date of text: 13 June 2013 Regulation of Salt Iodization Date of text: 28 February 2011 Regulation on Grains and Rooted Crops Reserve Date of text: 17 October 2009 Regulation on Breastfeeding Support Date of text: 16 February 2009
2.	Australia	Food Standards Australia New Zealand Act 1991 Date of original text: 27 June 1991 (22 May 2018) Wheat Export Marketing Amendment Act 2012 (No. 170, 2012) Date of text: 03 December 2012 Wine Australia Act 2013 Date of original text: 16 March 2011 (12 March 2019) Wheat Export Marketing (Repeal and Consequential Amendments) Act 2008 (No. 66 of 2008) Date of text: 30 June 2008 Australia New Zealand Food Authority Amendment Act 2001 (No. 81 of 1001) Date of original text: 10 July 2001 (02 August 2002)
3.	Bangladesh	Food Safety Act, 2013 Date of text: 10 October 2013 Animal Slaughter & Meat Quality Control Act, 2011 (Act No. 16 of 2011) Date of text: 20 September 2010

(continued)

[3] This list of legislation is attached in reference to Sect. 6.2 of Chap. 6.

(continued)

		Formalin Control Act, 2015 (Law No. 5 of 2015) Date of text: 18 February 2015 Vitamin A Enrichment in Edible Oil Act, 2013 (Act No. 65 of 2013) Date of text: 27 November 2013 Geographical Indication of Goods (Registration and Protection) Act, 2013 (Act No. 54 of 2013) Date of text: 10 November 2013
4.	Bhutan	Food Act, 2005 Date of text: 30 November 2005 Food Rules and Regulations of Bhutan, 2017 Date of text: 05 July 2017 Slaughter and meat inspection 1981 (By-law No. 5) Date of text: 1981 Veterinary food and commodity inspection 1981 (By-law No. 6) Date of text: 1981 Centenary Farmers Market Management Rules and Regulations, 2009 Date of text: 2009
5.	Canada	A Healthy Environment and a Healthy Economy Date of text: 2020 Food Policy for Canada: Everyone at the Table Date of text: 2019 Safe Food for Canadians Act (S.C. 2012, c. 24) Date of original text: 2012 (19 June 2014) Food and Drugs Act (R.S. 1985, c. F-27) Date of original text: 1985 (25 March 2020) Canadian Food Inspection Agency Act (S.C. 1997, c. 6) Date of original text: 20 March 1997 (17 June 2019) Date of original text: 2013 (01 November 2013 Safe Food for Canadians Regulations (SOR/2018-108) Date of original text: 10 September 2018 (17 June 2019) Safe Drinking Water for First Nations Act
6.	China	Fourteenth Five-Year Plan for Cold Chain Logistics Development Date of text: 26 November 2021 Anti-food Waste Law of the People's Republic of China
		Date of text: 29 April 2021 Food Safety Law (2015) Date of original text: 28 February 2009 (24 April 2015) Decree No. 49 of President laying down the Law of the People's Republic of China on Quality and Safety of Agricultural Products Date of text: 29 April 2006 Food Hygiene Law of the People's Republic of China Date of text: 30 October 1995 Law of the People's Republic of China on Protecting Consumers' Rights and Interests Date of text: 31 October 1993

(continued)

(continued)

7	Ethiopia	Food and Medicine Administration Proclamation No. 1112/2019 Date of text: 28 February 2019 Food, Medicine and Health Care Administration and Control Proclamation No. 661/2009 Date of text: 13 January 2010 Coffee Marketing and Quality Control Proclamation No. 1051/2017 Date of text: 01 July 2017 Quality and Standards Authority of Ethiopia Establishment Proclamation (No. 102/1998) Date of text: 03 March 1998
8.	India	National Food Security Act, 2013 Date of text: 10 September 2013 Food Safety and Standards Act, 2006 Date of text: 23 August 2006 Food Safety and Standards (Recovery and Distribution of Surplus Food) Regulations, 2019. Date of text: 26 July 2019
		Food Safety and Standards (Food Products Standards and Food Additives) Regulations, 2011 Date of text: 01 August 2011 Agricultural and Processed Food Products Export Development Authority (Amendment) Act, 2009 Date of text: 31 October 2008
9.	Japan	Food Safety Basic Act (Act No. 48 of 2003) Date of text: 2003 Food, Agriculture and Rural Areas Basic Act (Act No. 106 of 1999) Date of original text: 16 July 1999 (29 July 2005)
		Act on Japanese Agricultural Standards (Act No. 175 of 1950) Date of original text: 11 May 1950 (2017) Food Sanitation Act (Act No. 233 of 1947) Date of original text: 24 December 1947 (2006) Food Labelling Act (Act No. 70 of 2013) Date of original text: 28 June 2013 (2014)
10.	Korea	Framework Act on Agriculture, Fisheries, Farming and Fishing Villages and Food Industry Date of original text: 27 May 2009 (09 March 2011) National Nutrition Management Act Date of original text: 26 March 2010 (29 December 2015) Framework Act on Food Safety Date of original text: 13 June 2008 (02 December 2016) Special Act on Imported Food Safety Control Date of text: 03 February 2015 Act on Tea Industry Development and Tea Culture Promotion Date of text: 20 January 2015
11.	Netherlands	Emergency Food Supply Act Date of original text: 12 December 1962 (01 January 2015) Decree on the preparation and treatment of food made under the Merchandise Act Date of original text: 10 December 1992 (03 March 2016)

(continued)

(continued)

		Act No. 247 of 2006 concerning rules relative to responsibilities of competent authorities for purposes of enforcement of legislation regarding consumer protection (Act Enforcement Consumer Protection) Date of original text: 20 November 2006 (19 March 2012) Act No. 247 of 2006 concerning independent risk analysis by the Authority for Food and Consumer Goods (Act Independent Risk Analysis Authority for Food and Consumer Goods) Date of original text: 26 April 2006 (19 March 2012) Emergencies Coordination Act Date of original text: 03 April 1996 (01 January 2018)
12.	Nigeria	Agricultural Sector Food Security and Nutrition Strategy, 2016- 2025 Date of text: 01 May 2017 Food Safety and Quality Bill, 2019 Date of text: 2019 Counterfeit and Fake Drugs and Unwholesome Processed Food Act Date of original text: 10 May 1999 (12 February 2013) National Agency for Food and Drug Administration and Control Act Date of original text: 27 January 1993 (2004) Food, Drugs and Related Products (Registration, etc.) Act Date of original text: 27 January 1993 (14 February 2013) Consumer Protection Council Act Date of original text: 23 November 1992 (12 February 2013)
13.	Pakistan	Pakistan Halal Authority Act, 2016 (No. VIII) Date of text: 27 February 2016 Protection of Breast-feeding and Child Nutrition Ordinance, 2002 Date of text: 26 October 2002 Pakistan Fish Inspection and Quality Control Act (No. XXXV of 1997) Date of text: 06 December 1997 Cantonments Pure Food Act (Act No. XVI of 1966). Date of text: 19 July 1966
		Pakistan Nuclear Safety and Radiation Protection (Treatment of Food by Ionizing Radiation) Regulations, 1996 Date of text: 05 March 1996
14.	Singapore	Singapore Food Agency Act 2019 (No. 11 of 2019) Date of text: 07 March 2019 Sale of Food Act (Chapter 283) Date of original text: 01 May 1973 (01 April 2019) Health Promotion Board Act (Chapter 122B) Date of original text: 30 March 2001 (31 July 2002) Wholesome Meat and Fish Date of original text: 10 December 1999 (01 April 2003)
		Control of Essential Supplies Act, (Chapter 55);revised version 1985 Date of original text: 14 December 1973 (30 November 1984)
15.	Somalia	Strategic Plan of the MINISTRY OF AGRICULTURE 2016–2020 Date of text: 01 June 2016 National Dairy Act Date of text: 2015

(continued)

(continued)

16.	Sudan	National Biosafety Law No.15 of 2010 Date of text: 29 June 2010 Presidential Decree No.287 of 2015 to form the Higher Council for Food Security and Nutrition Date of text: 2015 Ordinance of the Council of Ministers No. 5 of 2007 establishing the General Authority for Goods Strategic Reserve Date of text: 18 February 2007 Law No. 569 of 1973 on food control Date of text: 18 December 1973 Food Control Regulation (inspection, sampling and analysis) No. 4 of 1980 Date of text: 1980
17.	Syria	National Programme for Food Security in the Syrian Arab Republic Date of text: 01 August 2010 Law No.19 of 2008 to ensure the safety, health and quality of food Date of text: 28 October 2008
		Legislative Decree 8/2021 regarding the Consumer Protection Law Date of text: 12 April 2021 Grains Law No.11 of 2019 Date of text: 22 April 2019 Law No. 23 of 2008 regulating the extraction, production and marketing of the salt in Syria Date of text: 18 November 2008 Resolution No. 1186 of 2008 regulating the transport and transfer of wheat and flour between the Syrian governorates (provinces) Date of text: 22 April 2008
18.	Tanzania	Tanzania Food, Drugs and Cosmetics Act, 2003 (No. 1 of 2003) Date of text: 10 February 2003 Food Security Act, 1991 (Act No. 10 of 1991) Date of text: 05 August 1991 Meat Industry Act, 2006 (No. 10 of 2006) Date of text: 2006 Dairy Industry Act, 2004 (No. 8 of 2004) Date of text: 04 June 2004 Kariakoo Market Corporation Act, 1974 (No. 36 of 1974) Date of text: 30 October 1974
19.	United Kingdom	Food Standards Act 1999 (Chapter 28) Date of original text: 11 November 1999 (01 September 2018) Coronavirus Act 2020 (Chapter 7) Date of text: 25 March 2020 Groceries Code Adjudicator Act 2013 (Chapter 19) Date of original text: 25 April 2013 (01 April 2014) Consumers, Estate Agents and Redress Act 2007 (Chapter 17) Date of text: 19 July 2007 School Standards and Framework Act 1998 (Chapter 13) Date of original text: 24 July 1998 (2013)

(continued)

(continued)

20.	United States	United States Health Security National Action Plan Date of text: 01 October 2018 Global Food Security Act of 2016 Date of text: 06 July 2016 Food Safety Modernization Act (P.L. 111–353)
		Date of text: 04 January 2011 FDA Food Safety Modernization Act (21 U.S.C. 2201-2252) Date of original text: 04 January 2011 (10 December 2015) Food Safety (21 U.S.C. 2101-2110). Date of original text: 27 September 2007 (19 September 2016) Organic Foods Production Act of 1990 Date of original text: 28 December 1990 (10 November 2005)

MIX
Papier aus verantwortungsvollen Quellen
Paper from responsible sources
FSC® C105338

If you have any concerns about our products,
you can contact us on
ProductSafety@springernature.com

In case Publisher is established outside the EU,
the EU authorized representative is:
Springer Nature Customer Service Center GmbH
Europaplatz 3, 69115 Heidelberg, Germany

Printed by Libri Plureos GmbH
in Hamburg, Germany